*Training Circular
No. 3-22.90 (FM 3-22.90)

Headquarters
Department of the Army
Washington, DC, 17 March 2017

Mortars

Contents

		Page
	PREFACE	viii
Chapter 1	**Introduction**	**1-1**
	History of Mortars	1-1
	Modern Use	1-1
	Indirect Fire Team	1-2
	Safety Procedures	1-3
	Loading and Firing	1-14
Chapter 2	**Mortar Equipment and Employment**	**2-1**
	Boresighting	2-1
	M2 Compass	2-7
	M2 and M2A2 Aiming Circles	2-10
	Sight Units	2-11
	Other Equipment	2-16
	Laying of the Section	2-20
Chapter 3	**60-mm Mortar**	**3-1**
	Organization and Duties	3-1
	Components	3-1
	Operation	3-5
	Modes of Firing the 60-mm Mortar	3-11
Chapter 4	**81-mm Mortar**	**4-1**
	Organization and Duties	4-1
	Components	4-2
	Operation	4-5

Distribution Restriction: Approved for public release; distribution will be unlimited.

*This publication supersedes FM 3-22.90, 7 December 2007

Contents

Chapter 5	**120-mm Mortars**	**5-1**
	Organization and Duties	5-1
	Components	5-2
	M120 and M121 Operations	5-6
	M326 120-mm Mortar Stowage Kit Operations	5-14
	M1064A3 Mortar Carrier Operations	5-33
	M1129A1 Stryker Mortar Carrier Vehicle Operations	5-41
Chapter 6	**Mortar Fire Control System**	**6-1**
	MFCS Capabilities	6-1
	Mortar Fire Control System Mounted	6-1
	Mortar Fire Control System-Dismounted	6-8
	Morter Fire Control System Operations	6-15
Chapter 7	**Fire Without a Fire Direction Center**	**7-1**
	Direct Lay and Direct Alignment	7-1
	Firing Data	7-2
	Direct Lay Method	7-3
	Direct Alignment Method	7-5
	Fire Commands	7-5
	Modified Fire Commands	7-6
	Observer Corrections	7-8
	Adjustment Methods	7-9
	Squad Conduct of Fire	7-15
	Use of Smoke and Illumination	7-16
	Search or Traverse	7-16
Chapter 8	**Gunner's Examination**	**8-1**
	Preparation and Instruction	8-1
	Ground-Mounted Mortar Examination	8-5
	M121 Carrier-Mounted or the RMS6-L Stryker-Mounted Mortar Examination	8-17
Appendix A	**Ammunition**	**A-1**
Appendix B	**Training Devices**	**B-1**
Glossary	Glossary-	1
References	References-	1
Index	Index-	1

Contents

Figures

Figure 1-1. Indirect fire team 1-2
Figure 1-2. Completed safety record or card 1-9
Figure 1-3. Safety diagrams for M821 HE and M853A1 illumination
　　　　　　 rounds 1-10
Figure 1-4. Safety "T" for M821 HE 1-12
Figure 1-5. Safety "T" for M853A1 illumination round 1-13
Figure 1-6. Arm-and-hand signals for ready, fire, and cease-fire 1-20
Figure 2-1. M45 boresight 2-2
Figure 2-2. M115 boresight 2-3
Figure 2-3. Gunner's quadrant sight 2-6
Figure 2-4. M2 compass (top view) 2-7
Figure 2-5. M2 compass (side view) 2-9
Figure 2-6. M2 compass (user's view) 2-9
Figure 2-7. M67 sight unit 2-11
Figure 2-8. Warning label for tritium gas (H^3) 2-16
Figure 2-9. M14 and M1A2 aiming posts 2-17
Figure 2-10. M58 and M59 aiming post lights 2-17
Figure 2-11. Parallel sheaf 2-20
Figure 2-12. Calibration for deflection using the angle method 2-21
Figure 2-13. Calibration for deflection using the distant aiming point
　　　　　　　method 2-22
Figure 2-14. Mortar laid parallel with sights 2-23
Figure 2-15. Sighting on the mortar sight 2-24
Figure 2-16. Arm-and-hand signals used in placement of aiming
　　　　　　　posts 2-26
Figure 3-1. M224/M224A1 60-mm mortar 3-2
Figure 3-2. M225/M225A1 cannon assembly 3-3
Figure 3-3. M170/M170A1 bipod assembly 3-4
Figure 3-4. M7/M7A1 baseplate 3-5
Figure 3-5. M8/M8A1 baseplate 3-5
Figure 3-6. Large deflection and elevation changes 3-9
Figure 3-7. Compensated sight picture 3-10
Figure 3-8. M224/M224A1 range indicator assembly 3-14

17 March 2017　　　　　　　　TC 3-22.90　　　　　　　　iii

Contents

Figure 3-9. M224A1 Selector knob...3-15
Figure 4-1. Position of squad members 4-2
Figure 4-2. M252/M252A1 81-mm mortar...4-3
Figure 4-3. M253 cannon assembly 4-3
Figure 4-4. M177/M177A1 mount in folded position..4-4
Figure 4-5. M3A1/A2 baseplate...4-5
Figure 4-6. Arrangement of equipment 4-6
Figure 4-7. Baseplate placed against baseplate stake 4-7
Figure 5-1. Position of squad members 5-2
Figure 5-2. 120-mm mortar..5-3
Figure 5-3. M298 cannon assembly 5-4
Figure 5-4. M191 bipod assembly (carrier- or ground-mounted) 5-5
Figure 5-5. M9A1 baseplate...5-6
Figure 5-6. M326 mortar stowage kit 5-15
Figure 5-7. Ammunition racks mounted on the M1101 trailer 5-16
Figure 5-8. Ammunition racks unmounted 5-16
Figure 5-9. Squad member positions 5-17
Figure 5-10. Lift drive selector...5-20
Figure 5-11. Manual winch drum and guide hole 5-21
Figure 5-12. Support strut eye hook...5-21
Figure 5-13. Barrel clamp...5-22
Figure 5-14. Winch strap aligned with the manual winch standoff.............5-23
Figure 5-15. Gunner guide to apex 5-24
Figure 5-16A. Support strut and baseplate brackets....................................5-27
Figure 5-16B. Tube clamp..5-28
Figure 5-17. Support strut tube clamp...5-28
Figure 5-18. Mortar ready for lifting 5-29
Figure 5-19. Lift hook in position 5-29
Figure 5-20. Lift drive selector...5-31
Figure 5-21. Winch strap guide hole 5-31
Figure 5-22. Manual winch standoff..5-32
Figure 5-23A. M1064A3 mortar carrier, front and side.................................5-33
Figure 5-23B. M1064A3 mortar carrier, rear view...5-33
Figure 5-24. Stryker mortar carrier vehicle..5-42
Figure 5-25. Squad configuration 5-43

Contents

Figure 5-26. RMS6-L 120-mm mortar 5-44
Figure 5-27. Mortar doors 5-45
Figure 5-28. Horizontal ammunition rack.. 5-47
Figure 5-29. Right side ammunition rack.. 5-50
Figure 5-30. Example distant aiming point for boresight 5-51
Figure 5-31. M67 sight unit... 5-52
Figure 5-32. Cant correction (cross-level knob) 5-52
Figure 5-33. Blast attenuator device 5-53
Figure 5-34. M45 boresight on cannon... 5-53
Figure 5-35. M45 boresight elevation bubble 5-54
Figure 6-1. Mortar Fire Control System 6-3
Figure 6-2. Fire control computer 6-4
Figure 6-3. Power distribution assembly.. 6-5
Figure 6-4. Pointing device 6-6
Figure 6-5. Gunner's display 6-7
Figure 6-6. Driver's display 6-7
Figure 6-7. Vehicle motion sensor.. 6-8
Figure 6-8. M120 with M150 mounted 6-9
Figure 6-9. Fire control computer 6-10
Figure 6-10. Enhanced power distribution assembly........................... 6-11
Figure 6-11. Pointing device mounting assembly 6-12
Figure 6-12. Gunner's display.. 6-13
Figure 6-13. Portable universal battery supply 6-14
Figure 6-14. Electronics rack.. 6-14
Figure 6-15. Pointing device status screen.. 6-15
Figure 6-16. Boresight screen 6-17
Figure 6-17. Navigation/emplacement screen..................................... 6-18
Figure 6-18. Driver's display showing steering directions, distance,
  and position.. 6-18
Figure 6-19. Driver's display showing arrival at the waypoint.............. 6-19
Figure 6-20. Navigation/emplacement screen: message transmittal
  and reception by the FDC 6-20
Figure 6-21. Navigation/emplacement screen: fire area...................... 6-21
Figure 6-23. Fire command screen 6-22
Figure 6-24. Subsequent fire command screen 6-23

17 March 2017 TC 3-22.90 v

Contents

Figure 6-25. End of mission screen 6-23
Figure 6-26. Not-in-mission screen 6-24
Figure 6-27. Fire command for an assigned final protective fire 6-25
Figure 6-28. End of mission 6-26
Figure 6-29. Fire the stored final protective fire 6-26
Figure 7-1. Graphical firing scale 7-3
Figure 7-2. Observer more than 100 meters from mortar but within
100 meters of G-T line 7-9
Figure 7-3. Bracketing method 7-10
Figure 7-4. Ladder method of fire adjustment 7-11
Figure 7-5. Adjustment of fire onto a new target with the observer
within 100 meters of the G-T line 7-14
Figure 7-6. Searching fire 7-18
Figure 7-7. Traversing fire 7-19
Figure 8-1. Sample of completed DA Form 5964 8-3
Figure 8-2. Equipment and personnel for the gunner's examination
(60-mm mortar) 8-7
Figure 8-3. Equipment and personnel for the gunner's examination
(M252/M252A1 81-mm mortar) 8-8
Figure 8-4. Equipment and personnel for the gunner's examination
(120-mm mortar with mortar stowage kit) 8-9
Figure A-1. Fin assembly A-2
Figure A-2. Stacked ammunition A-5
Figure A-3. M734 multioption fuze A-22
Figure A-4. M734 multioption fuze setting A-23
Figure A-5. M935 point-detonating fuze A-25
Figure A-6. M783 point-detonating and delay fuze A-27
Figure A-7. M745 point-detonating fuze A-28
Figure A-8. M776 mechanical time A-30
Figure A-9. Floating firing pin A-33

Tables

Table 1-1. M821 HE firing table 1-12
Table 1-2. M853A1 ILLUM firing table 1-12
Table 1-3. Sequence for transmittal of fire commands 1-17

Contents

Table 2-1. Tabulated boresight data for M45- and M115-series 2-1
Table 2-2. M67 sight unit equipment data 2-12
Table 6-1. Function keys 6-4
Table 7-1. Initial range change 7-10
Table 8-1. Conduct of ground-mounted gunner's examination 8-6
Table 8-2. Conduct of carrier-mounted gunner's examination 8-18
Table A-1. Mortar ammunition color codes A-4
Table A-2. M224/M224A1 60-mm mortar high-explosive rounds A-6
Table A-3. M224/M224A1 60-mm mortar illumination rounds A-9
Table A-4. Smoke, white phosphorus ammunition for the
            M224/M224A1 60-mm mortar A-10
Table A-5. M224/M224A1 60-mm mortar practice rounds A-11
Table A-6. M252/M252A1 81-mm mortar HE rounds A-12
Table A-7. M252/M252A1 81-mm mortar illumination rounds A-13
Table A-8. M252/M252A1 81-mm mortar smoke, white
            phosphorus, and red phosphorous rounds A-15
Table A-9. M252/M252A1 81-mm mortar training and practice
            rounds A-16
Table A-10. M120 and M121 120-mm mortar high-explosive
            rounds A-17
Table A-11. M120- and M121 120-mm mortar illumination rounds A-18
Table A-12. M120 and M121 120-mm mortar smoke, white
            phosphorus rounds A-19
Table A-13. M120 and M121 120-mm mortar practice round A-20

Preface

This publication prescribes guidance for leaders and members of mortar squads. It concerns mortar squad training and is used with the applicable technical manuals (TMs) and Army training programs. It presents practical solutions to assist in the timely delivery of accurate mortar fires, but does not discuss all possible situations. Local requirements may dictate minor variations from the methods and techniques described herein. However, principles should not be violated by modification of techniques and methods.

The scope of this publication includes mortar squad training at the squad level. The 60-milimeter (mm) mortar, M224/M224A1; 81-mm mortar, M252/M252A1; 120-mm mortars, M120/M121; and the Recoil Mortar System 6–Lightweight (RMS6-L) are discussed, and includes nomenclature, sighting, equipment, characteristics, capabilities, and ammunition. (Refer to ATTP 3-21.90 for information on the tactics, techniques, and procedures that mortar sections and platoons use to execute the combat mission.)

This publication prescribes DA Form 5964 *(Gunner's Examination Scorecard–Mortars)*.

Uniforms depicted in this manual were drawn without camouflage for clarity of the illustration.

Commanders, staffs, and subordinates ensure that their decisions and actions comply with applicable United States, international, and in some cases host-nation laws and regulations. Commanders at all levels ensure that their Soldiers operate according to the law of war and the rules of engagement. (See FM 27-10.)

This publication applies to the active Army, the Army National Guard, the Army Reserve, and the United States Marine Corps. Unless otherwise stated in the publication, masculine nouns and pronouns refer to both men and women.

TC 3-22.90 uses joint terms where applicable. Selected joint and Army terms and definitions appear in both the glossary and the text. Terms for which TC 3-22.90 is the proponent publication (the authority) are italicized in the text and marked with an asterisk (*) in the glossary. Terms and definitions for which TC 3-22.90 is the proponent publication are boldfaced in the text. For other definitions shown in the text, the term is italicized and the number of the proponent publication follows the definition.

The proponent of this publication is U.S. Army Maneuver Center of Excellence (MCoE). The preparing agency is the MCoE, Fort Benning, Georgia. You may submit comments and recommended changes in any of several ways—U.S. mail, e-mail, fax, or telephone—as long as you use or follow the format of DA Form 2028, *(Recommended Changes to Publications and Blank Forms)*. Contact information is as follows:

    E-mail: usarmy.benning.mcoe.mbx.doctrine@mail.mil
    Phone: COM 706-545-7114 or DSN 835-7114
    Fax: COM 706-545-8511 or DSN 835-8511
    U.S. mail: Commander, MCoE
                  Directorate of Training and Doctrine (DOTD)
                  Doctrine and Collective Training Division
                  ATTN: ATZB-TDD
                  Fort Benning, GA 31905-5410

Chapter 1

Introduction

Mortars are suppressive indirect fire weapons. They can be employed to neutralize, suppress, or destroy area or point targets, screen large areas with smoke, and provide illumination or coordinated high-explosive/illumination. The mortar platoon's mission is to provide close and immediate indirect fire support to all maneuver units on the battlefield.

HISTORY OF MORTARS

1-1. Since the 15th century when they first appeared, mortars have contributed greatly to the maneuver force. The first mortars were large and heavy, not becoming portable until around the time of the American Civil War; where mortars influenced operations and the outcomes of battles. Later improvements to mortars allowed for one Soldier to carry and operate the weapon. During the World War II Sicilian Campaign in 1943, 4.2-inch mortars stopped a number of enemy counterattacks and forced tanks to withdraw to a range where artillery and naval gunfire could destroy them. During urban fighting in France in 1944, U.S. units used 60-mm, 81-mm, and 4.2-inch mortars in house-to-house fighting. The Soldiers were able to fire over high buildings and into relatively narrow open spaces.

1-2. Mortars continue to improve and have been used in many conflicts including Korea, Vietnam, and Operations Enduring Freedom and Iraqi Freedom. Today, they provide indirect fire and are organic to the maneuver commander, which aids in the success of the unit's mission. Their ability to provide high angle fires is invaluable against dug-in enemy troops. Currently, there are three major calibers of mortars used by the Army; the 60-mm, 81-mm, and the 120-mm.

MODERN USE

1-3. Modern combat demands timely and accurate delivery of indirect fire to meet the needs of the supported unit. All indirect fire team members must be trained to quickly execute an effective fire mission. Mortar units must be prepared to accomplish multiple fire missions by providing an immediate, heavy volume of accurate fire for sustained periods.

1-4. For mortar fire to be effective, it must be dense and hit the target at the right time with the right projectile and fuze. The objective is to deliver a large volume of accurate and timely fire, with the time required to bring fires on the target kept to a minimum. A demoralizing effect on the enemy can be achieved by delivering the maximum number of rounds from all the mortars, in the shortest possible time.

Chapter 1

INDIRECT FIRE TEAM

1-5. Indirect fire procedure is a team effort. (See figure 1-1.) Since the mortar normally is fired from defilade, the indirect fire team gathers and applies the required data for firing. The team consists of a forward observer (FO), a fire direction center (FDC), and the mortar squad.

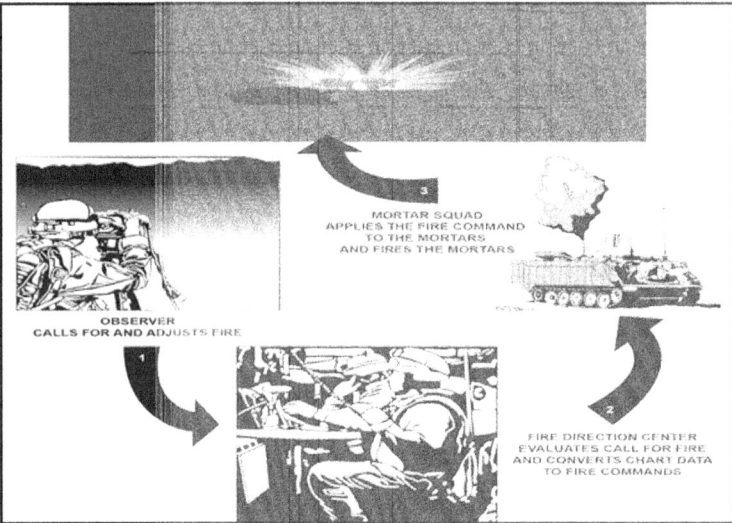

Figure 1-1. Indirect fire team

FORWARD OBSERVER

1-6. The team's mission is to provide accurate and timely response to the unit it supports. Effective communication is vital to the successful coordination of the indirect fire team's efforts.

1-7. The headquarters and headquarters battery of the supporting fires battalion provides the fire support teams (FISTs) to the maneuver battalions upon deployment. The FISTs typically move to and remain with their supported companies and platoons. The FOs job is to find and report the location of targets, and to request and adjust fire.

1-8. Each rifle company is supported by a 10-Soldier FIST section consisting of a lieutenant, a staff sergeant, a radio-telephone operator (RTO), and a driver with an armored personnel carrier in Armored brigade combat teams (ABCTs) or a high mobility multipurpose wheeled vehicle (HMMWV) in Infantry brigade combat teams at company headquarters. There are also six forward observers; one two-Soldier team for each Infantry platoon in the company.

1-9. Each ABCT or Stryker company and cavalry troop is supported by a four-Soldier FIST consisting of a lieutenant, staff sergeant, RTO, and a driver with an armored personnel carrier at company headquarters.

Introduction

FIRE DIRECTION CENTER

1-10. Fire direction includes the methods and techniques FDCs use to convert calls for fire (CFFs) into proper fire commands. The FDC is the element of the mortar platoon headquarters that controls the fire of a mortar section and relays information and intelligence from observers to higher headquarters. Tactical fire direction involves the FDC's control of mortars in the selection of targets, the designation of units to fire, and the allocation of ammunition for each mission.

1-11. The FDC has two computer personnel who convert the data in a call for fire (CFF) from the FO into firing data that can be applied to the mortars. Medium and heavy mortars also have a driver that acts as security/RTO for the FDC element.

MORTAR SQUAD

1-12. The mortar squad consists of three to five Soldiers. However, the number of Soldiers assigned to a squad varies depending on the unit's mission or personnel strength. The mortar squad has the following responsibilities:
- The squad leader is in charge of the squad and supervises the emplacement, laying, and firing of the mortar.
- The gunner manipulates the sight by placing the firing data on the sight and lays the mortar for deflection and elevation; he also manipulates the elevating gear handle and traversing the assembly wheel to align the mortar system back on target.
- The assistant gunner loads the mortar and assists the gunner in shifting the mortar. When faced with reduced manning or unavailability of an ammunition bearer, the assistant gunner may serve a dual role of both assistant gunner and ammunition bearer.
- The ammunition bearer or driver prepare the ammunition for firing and provide local security while rounds are being fired.

SAFETY PROCEDURES

1-13. Although safety is a command responsibility, each member of the mortar squad must know safety procedures and enforce them. This is enhanced with MFCS features, such as the ability to view safety data and situational awareness information, as well as inputting fire support coordination measures and view them using the digital plot feature. No matter how sophisticated the system, always check manually before firing.

1-14. Safety officers must help commanders meet the responsibility of enforcing safety procedures. The safety officer has two principal duties. First, he ensures that the section is properly laid so that, when rounds are fired, they land in the designated impact area. Second, he ensures that all safety precautions are observed at the firing point (FP).

Note. The safety officer must be an Infantry Mortar Leader's Course (IMLC) graduate, and should be a platoon leader, platoon sergeant, or section leader.

1-15. Training preparation involves two steps:
- 1. Conduct a training risk assessment.
- 2. Conduct an environmental risk assessment.

Chapter 1

CONDUCT RISK MANAGEMENT AND ASSESSMENT

1-16. The officer in charge (OIC) or noncommissioned officer in charge (NCOIC) conducts a training risk assessment. It is vital to identify unnecessary risks by comparing potential benefit to potential loss. The risk management (RM) process allows units to identify and control hazards, conserve combat power and resources, and complete the mission. This process is cyclic and continuous; it must be integrated into all phases of operations and training.

1-17. There are five steps to the RM process:
- Step 1- Identify hazards.
- Step 2- Assess hazards to determine risk.
- Step 3- Develop controls and make risk decisions.
- Step 4- Implement controls.
- Step 5- Supervise and evaluate.

Note. Risk decisions must be made at the appropriate level. (Refer to ATP 5-19 for more information.)

Assessment Steps

1-18. Steps 1 and 2 of RM are assessment steps—*risk assessment* is the identification and assessment of hazards. When identifying hazards (Step 1), leaders should consider—
- The lethality of the mortar and munitions used.
- The area in which training is to be conducted.
- How the addition of new elements impacts known hazards.

1-19. Once identified, a hazard is assessed to determine risk (Step 2) by considering the likelihood of its occurrence and the potential severity of injury without considering any control measures. When assessing hazards, leaders should consider the Soldiers' current state of training.

Management Steps

1-20. Steps 3 through 5 of RM are management steps

1-21. Leaders must apply two types of control measures (Step 3) to the mortar training risk assessments:
- Educational controls.
- Physical controls.

1-22. The unit commander's controls should be clear, concise, executable orders.

Note. Most vital to developing RM controls is mature, educated leadership.

1-23. **Educational controls** occur when adequate training takes place. They require the largest amount of planning and training time. Leaders implement educational controls using two sequential steps:

Step 1. Supervisors and instructors must be certified.

Step 2. Soldier training must be executed.

Introduction

1-24. **Physical controls** are the measures emplaced to reduce injuries. These include not only protective equipment, but certified personnel to supervise the training. Unrestrained physical controls are a hazard.

1-25. When leaders **implement the controls** (Step 4), they must match the controls to the Soldier's skill level. They also must enforce every control measure as a means of validating its adequacy.

1-26. **Supervising and evaluating** (Step 5) enables leaders to eliminate unnecessary risk and ineffective controls by identifying unexpected hazards and determining if the implemented controls reduced the residual risk without interfering with the training.

DUTIES BEFORE DEPARTING FOR RANGE

1-27. The safety officer must read and understand—
- Army Regulation (AR) 385-63.
- Post range and terrain regulations.
- The terrain request of the firing area to know safety limits and coordinates of firing positions.
- Appropriate Field Manuals (FMs), Training Circulars (TCs) and Technical Manuals (TMs) pertaining to weapons and ammunition to be fired. This also includes vehicle safety checks; and live-fire safety checks when mortar systems are mounted, according to the most current mortar system TMs.

DUTIES OF SUPERVISORY PERSONNEL

1-28. Supervisory personnel must know the immediate action to be taken for firing accidents. The following is a list of minimum actions that must be taken if an accident occurs:
- Administer first aid to injured personnel, and then call for medical assistance.
- If the ammunition or equipment presents further danger, move all personnel and equipment out of the area.
- Do not change or modify any settings or positions of the mortar until an investigation has been completed.
- Record the ammunition lot number involved in the accident or malfunction, and report it to the battalion ammunition officer. If a certain lot number is suspect, its use should be suspended by the platoon leader.

MORTAR RANGE SAFETY CHECKLIST

1-29. A mortar range safety checklist can be written for local use. The sections below are suggested checklists for the different stages of range operations. Leader checks are performed before, during, and after firing. (It can include three columns on the right, titled Yes, No, and Remarks.)

Leader Checks Before Firing

1-30. Before opening a range, the following questions need to be answered:
- Is a range log or journal maintained by the OIC?
- Is radio or telephone communication maintained with—

Chapter 1

- Range control?
- Unit operations staff officer (S-3)?
- Mortar squad?
- Forward observers?
- Road or barrier guards?
* Are the following required emergency personnel and equipment present on the range:
  - Properly briefed and qualified medical personnel?
  - A wheeled or tracked ambulance?
  - Firefighting equipment?
* Are the following range controls and warning devices available, readily visible, and in use during the firing exercise:
  - Barrier/road guards briefed and in position?
  - Road barriers in position?
  - Red range flag in position?
  - Blinking red lights for night firing?
  - Signs warning trespassers to beware of explosive hazards and not to remove duds or ammunition components from ranges?
  - Noise hazard warning signs?
* Are current copies of the following documents available and complied with:
  - AR 385-63/MCO3570.1C?
  - Technical and field manuals pertinent to the mortar in use?
  - Appropriate firing tables?
  - Installation range regulations?
* Are the following personal safety devices and equipment available and in use:
  - Helmets?
  - Protective earplugs?
  - Protective earmuffs (for double hearing protection)?
* Is the ammunition the correct caliber, type, and quantity required for the day's firing?
* Are the rounds, fuzes, and charges—
  - Stored in a location to minimize possible ignition or detonation?
  - Covered to protect them from moisture and direct sunlight?
  - Stacked on dunnage to keep them clear of the ground?
  - Strictly accounted for by lot number?
  - Exposed only immediately before firing?
  - Stored separately from ammunition and protected from ignition?
* Has a supervised inspection of previously handled and repackaged ammunition been conducted to ensure that all propellant charges are of the same LOT number and they are not excessively loose, damaged, or missing? Has each safety non-commissioned officer (NCO) and gunner received a safety "T" depicting:
  - Left and right limits (azimuth and deflection)?
  - Minimum and maximum elevation and charge?
  - Minimum time setting for fuzes?
* Has the range safety officer verified the following:

Introduction

- The mortar safety card applies to the unit and exercise?
- The firing position is correct and applies to the safety card, and the base mortar is within 100 meters of the surveyed firing point?
- Boresighting and aiming circle declination are correct?
- The plotting board, MFCS, or lightweight handheld mortar ballistic computer (LHMBC) is correct?
- The forward observer has been briefed on the firing exercise and knows the limits of the safety fan?
- The lay of each mortar is correct?
- The safety stakes (if used) are placed along the right and left limits?
- All personnel at the firing position have been briefed on safety misfire procedures?
- If the safety card specified overhead fire, firing is according to AR 385-63/MCO3570.1C?
- Is the borescope and pullover complete with a valid DA Form 2408-4 (*Weapon Record Data*) and updated on TULSA

Note. Firing mortars over the heads of unprotected troops by Marine Corps units is not authorized, and Army units require commander's approval. (Refer to DA Pam 385-63 for more information.) This restriction applies only during peacetime; it does not apply during combat. However, in a real-world environment it is vital that unit commanders ensure that there is no civilian populace along the gun-target (G-T) line. It is the unit commander's responsibility to ensure that there is no civilian habitation or traffic (vehicular or by foot) along the G-T line and clearance of the G-T line must be verified physically before a live-fire exercise may be performed.

- The mortars are safe to fire by checking:
  - Safety checks for each specific mortar.
  - The areas immediately forward of the gun line is clear of personnel
  - Check mask and overhead clearance.
  - Weapons and ammunition.
  - Properly seated sights on weapons.
  - All sight units have been boresighted properly.
  - The lights on the sights and aiming stakes (for night firing).
- The OIC is informed that the range is cleared to fire and that range control has placed it in a hot or wet status.

Note. Range status terms are range control or unit standard operating procedure (SOP)-driven.

Leader Checks

1-31. While operating a range, the following questions need to be answered:
- Are unit personnel adhering to the safety regulations?
- Each charge, elevation, and deflection setting is checked before firing?
- The squad leader declares the mortar safe to fire before announcing, "HANG IT."

Chapter 1

- Ensure all gun settings remain at last data announced until a subsequent fire command is issued by the FDC?
- Ammunition lots kept separate to avoid the firing of mixed lots?

1-32. When firing on a range is complete, the following considerations are taken:

- Have the gunners and safety NCO verified that no loose propellants are mixed with the empty containers?
- Has the safety NCO disposed of the unused propellants?
- Has the unused ammunition been inventoried and repacked properly?
- Have the proper entries been made on DA Form 2408-4?
- Has the OIC or safety officer notified range control of range status and other required information?
- Has a thorough range police been conducted?

SAFETY CARD/RECORD

1-33. The FDC receives a copy of the safety card/record from the OIC or range control—depending on local regulation—before allowing firing to begin. A safety card or record should be prepared and approved for each firing position and type of ammunition used. The form of the card depends upon local regulations (training list, overlay, range bulletin). Even without a prescribed format, it should contain the following:

- Unit firing or problem number.
- Type of weapon and fire.
- Authorized projectile, fuze, and charge zone.
- Grid of the platoon center.
- Azimuth of left and right limits.
- Minimum and maximum ranges and elevations.
- Any special instructions to allow for varying limits on special ammunition or situations.

SAFETY DIAGRAM AND SAFETY "T"

1-34. On receipt of the safety card or safety record, the FDC constructs a safety diagram. The safety diagram is a graphic portrayal of the data on the safety card and does not need to be drawn to scale, but must accurately list the sight settings that delineate the impact area. The safety "T" serves as a convenient means of checking the commands announced to the squad against those commands that represent the safety limits.

1-35. The diagram shows the right and left limits, and deflections corresponding to those limits; the maximum and minimum elevations; and the minimum fuze settings (when applicable) for each charge to be fired. The diagram also shows the minimum and maximum range lines, the left and right azimuth limits, the deflections corresponding to the azimuth limits, and the direction and mounting azimuth on which the guns are laid. The safety diagram must show only necessary information.

1-36. To accurately complete a safety diagram, the FDC must use the information supplied by range control or, in the example in figure 1-2, the safety card/record to:

- Enter the known data (supplied from the safety card) on the safety diagram.
- Determine the direction of fire or the center of the sector.
- Determine the mounting azimuth.

Introduction

- Determine mils left and right deviations of the mounting azimuth.
- Enter the referred deflection.
- Determine deflections to left and right limits.
- Determine minimum and maximum charges and elevations.
- If illumination (ILLUM) is to be used, determine, from the appropriate firing tables, the minimum and maximum charges and ranges to burst and impact for the canister. The minimum range is used to determine the minimum charge and range to burst. The maximum range is used to determine the maximum charge and range to impact.

1-37. A sample of the preparation of a safety diagram and safety "T" follows. An 81-mm mortar section is firing at FP 52 with M821 HE (high explosive) and M853A1 ILLUM. The safety officer receives the safety record/card from either range control or the OIC (according to local range regulations).

ARTILLERY/MORTAR SAFETY RECORD							
DATE: 10 December 2015								
FIRING POINT: FP 52				WEAPONS: M252, M224, M120/121			
COORDINATES: GL12345678			ELEV		FUZE		
Weapon Projectile	Left Limit Mils	Right Limit Mils	Minimum Range Meters	Maximum Range Meters	Minimum Charge	Maximum Charge	Maximum Ordnance Meters or Feet
M821HE	0500	0920	300	4000	0	4	NA
M853A1 IL	0500	0920	300	4000	0	4	NA
LEGEND ELEV ELEVATION FP FIRING POINT HE HIGH EXPLOSIVE IL ILLUMINATION							

Figure 1-2. Completed safety record or card

1-38. The FDC calculates the data for the safety diagram and then places the data for each type of ammunition on separate safety diagrams. (See figure 1-3 on page 1-10.)

Chapter 1

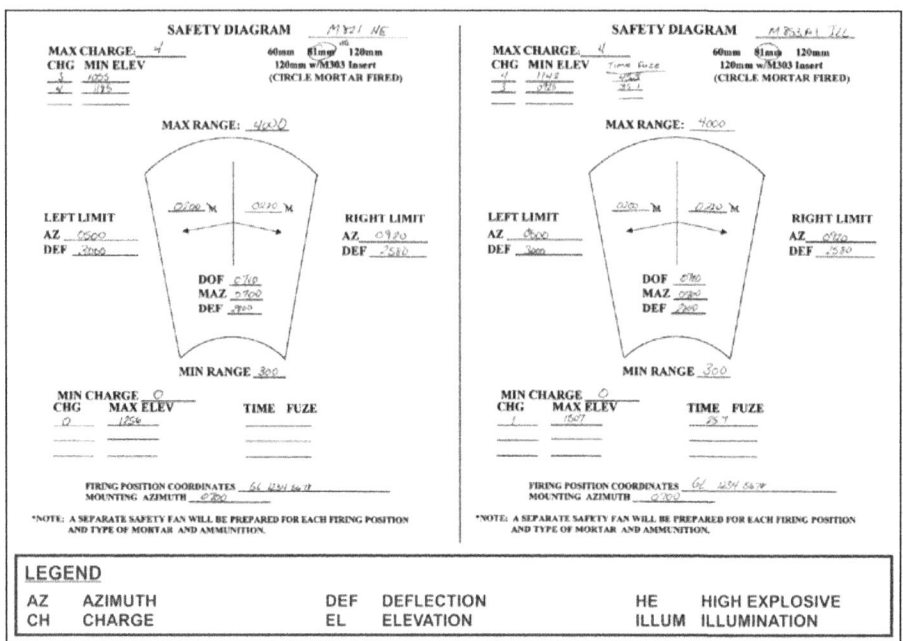

Figure 1-3. Safety diagrams for M821 HE and M853A1 illumination rounds

Calculate the Direction of Fire (Left Limit is Smaller Than Right Limit)

1-39. Subtract the left limit azimuth from the right limit azimuth. Divide the result by two, and add that number to the left limit azimuth.

SAMPLE	
RIGHT LIMIT	+ 0920
MINUS LEFT LIMIT	− 0500
SUM	420
SUM ÷ 2 =	210
LEFT LIMIT	+ 500
TOTAL	710

1-40. The answer, 710, is the direction of fire or the azimuth center of sector. (To calculate the direction of fire if the left limit is larger than the right limit, add 6400 to the right limit, and use the same calculations as above. If the final answer is more than 6400, subtract 6400 to get the direction of fire.)

Round Off the Direction of Fire or Mounting Azimuth and Enter the Referred Deflection

1-41. For all mortars using the plotting board, round off to nearest 50 mils. In this sample, the safety officer rounds 0710 to 0700.

1-42. The section leader provides the referred deflection. It can be any number, but 2800 is used, normally.

Determine Deflection for the Left and Right Limits

1-43. Determine the number of mils from the mounting azimuth to the left limit.

EXAMPLE	
MOUNTING AZIMUTH	0700
LEFT LIMIT	-0500
MILS TO LEFT LIMIT	0200

1-44. Using the left add/right subtract (LARS) rule for referred deflection, calculate the left limit deflection.

EXAMPLE	
CENTER OF SECTOR REFERRED DEFLECTION	2800
MILS TO LEFT LIMIT	+ 0200
LEFT LIMIT DEFLECTION	3000

1-45. Determine the number of mils from the mounting azimuth to the right limit.

EXAMPLE	
RIGHT LIMIT	0920
MOUNTING AZIMUTH	- 0700
MILS TO RIGHT LIMIT	0220

1-46. Using the LARS rule for referred deflection, calculate the right limit deflection.

EXAMPLE	
CENTER OF SECTOR REFERRED DEFLECTION	2800
MILS TO RIGHT LIMIT	- 0700
RIGHT LIMIT DEFLECTION	2100

Determine Minimum and Maximum Charges and Elevations

1-47. Use the firing tables (FTs) for the mortar being fired to determine minimum and maximum charges and elevations. In the sample, the maximum range is 4000 meters and the minimum range is 300 meters.

1-48. Using FT-81-AR-2 for the **M821 HE** cartridge, the maximum and minimum charges and elevations are detailed in table 1-1.

Table 1-1. M821 HE firing table

Maximum or Minimum	Charge	Elevation in mils
Maximum	4	1185
Maximum	3	1055
Minimum	0	1256

1-49. Using FT-81-AR-2 for the **M853A1 ILLUM** cartridge, the maximum and minimum charges, elevations, and time settings are detailed in table 1-2.

Table 1-2. M853A1 ILLUM firing table

Maximum or Minimum	Charge	Elevation in Mils	Time Setting in Seconds
Maximum	4	1142	45.5
Maximum	3	0925	35.1
Minimum	1	1507	25.7

1-50. A safety "T" is made for each type of cartridge being fired. The requisite data from the safety diagram is transcribed onto the safety "T" (see figure 1-4 and 1-5), and each gun has a copy.

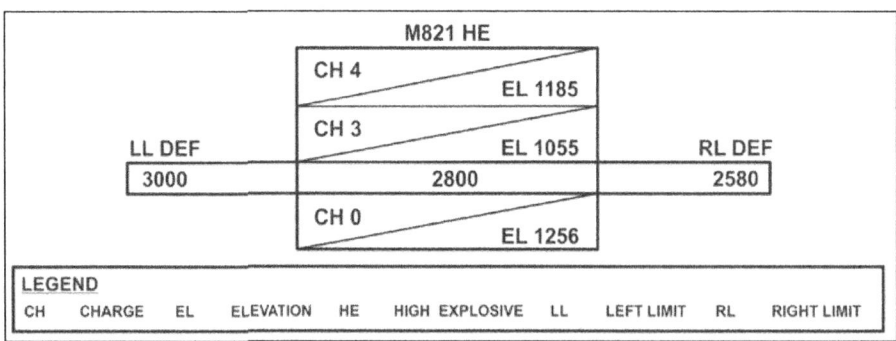

Figure 1-4. Safety "T" for M821 HE

Introduction

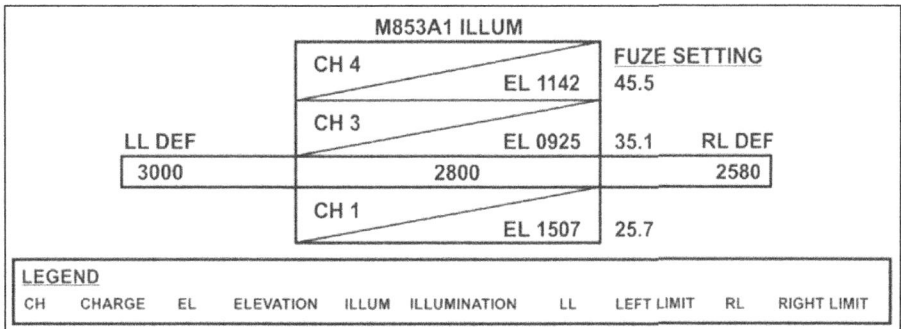

Figure 1-5. Safety "T" for M853A1 illumination round

SURFACE DANGER ZONES

1-51. The surface danger zone (SDZ) is the ground and airspace designated within the training complex for vertical and lateral containment of projectiles, fragments, debris, and components resulting from the firing, launching, or detonation of weapon systems

1-52. SDZs usually are pie-shaped, with the target area in the center surrounded by concentric zones. Components include the firing position, impact areas, and buffer zones.

Firing Positions

1-53. The firing position is the point or location at which the mortar is placed for firing. Mortar firing positions are selected based on: mission accomplishment (the most important factor), tactical situation, range criteria, target area coverage, survivability, overhead and mask clearance, surface conditions, communications, and routes.

1-54. With the introduction of automated fire direction systems, mortar leaders are no longer limited to positions that produce a parallel sheaf, provide a set pattern of rounds on target, or positions where all the mortars have to have line of sight to an aiming circle. When automated systems are not available, a mortar leader has to establish a firing position that enhances the ability to control fires with the M16-/M19-plotting board.

Impact and Target Areas

1-55. The impact area is the ground and associated airspace within the training complex used to contain fired or launched ammunition and explosives, and the resulting fragments, debris, and components from various weapon systems. Indirect fire weapon system impact areas include probable error for range and deflection.

1-56. The target area contains the targets.

Dispersion Area and Buffer Zones

1-57. By keeping the rounds inside the impact area, the dispersion area accounts for human error, gun or cannon tube wear, propellant temperature, and other various factors. Its width is based on

Chapter 1

the probable range and deviation errors for the maximum range of the charge permitted. The probable errors are described as follows:
- The left and right dimensions are eight probable deflection errors (PE$_D$) from the left and right limits of the target area.
- The far dimensions are eight probable range errors (PE$_R$) from the far edge of the target area.

1-58. Buffer zones, or secondary danger areas, contain the fragments, debris, and components from frangible or explosive projectiles and warheads functioning on the edge of the target area. A buffer zone has two parts: Area A and Area B.

Area A

1-59. Area A is the secondary danger area (buffer zone) that laterally parallels the impact area and dispersion area and contains fragments, debris, and components from frangible or explosive projectiles and warheads functioning on the right or left edge of the impact area or ricochet area. It starts at a 25-degree angle from the impact area. (It increases to 70 degrees for ranges at and beyond 600 meters for 60-mm; 940 meters for 81-mm; and 1415 meters for 120-mm mortars). The width of Area A depends on the type of mortar. The standard is as follows:
- A 250-meter width for 60-mm mortars.
- A 400-meter width for 81-mm mortars.
- A 600-meter width for 120-mm mortars.

Area B

1-60. Area B is the secondary danger area (buffer zone) on the downrange (far) side of the impact area and contains fragments, debris, and components from frangible or exploding projectiles and warheads functioning on the far edge of the impact area and Area A. The width of Area B depends on the type of mortar. The standard is as follows:
- A 300-meter width for 60-mm mortars.
- A 400-meter width for 81-mm mortars.
- A 600-meter width for 120-mm mortars.

1-61. SDZs usually exist for approved FPs and can be found at range control or in the range book for the FP. If the SDZ has to be constructed, personnel should refer to DA PAM 385-63 for more information.

LOADING AND FIRING

1-62. Upon receiving a fire command from the section leader, the entire mortar section repeats each element of it. The gunner places the firing data on the sight and, helped by the assistant gunner, lays the mortar. The ammunition bearer repeats the charge element when announced by the gunner and prepares the round with that charge.

1-63. If a fuze setting is announced, the ammunition bearer also repeats the setting and places it on the fuze. The ammunition bearer completes the preparation of the cartridge to include safety checks, removing charges, and removing any safety wires. The squad leader, for safety purposes, spot-checks the ammunition and the data on the sight and the lay of the mortar. Then commands, "FIRE."

Introduction

Note. When faced with a reduced mortar squad, on the 81-mm or 120-mm mortars (such as an ammunition bearer is not available), the assistant gunner must assume all those duties associated with the ammunition bearer plus normal duties. This increases the need for cross-talk among the squad to receive confirmation and acknowledgment from the squad leader that all data, specifically data pertaining to the mortar rounds, is correct and in concurrence with the FDC order or firing commands.

FIRING OF THE GROUND-MOUNTED MORTAR

1-64. The squad fires the mortar as follows:

- To settle the baseplate, the gunner removes the sight and is careful not to disturb the lay of the mortar. He continues to remove the sight until the baseplate assembly is settled and there is no danger of the sight becoming damaged from the recoil of the mortar. The bipod assembly can slide up the barrel when the gunner fires the M252/M252A1 81-mm mortar or the M120 120-mm mortar. The gunner ensures the bipod assembly is reset properly against the lower stop band between settling rounds. The gunner must not try to place the sight back into position until the baseplate settles.
- The FDC issues a fire command to the squad leader.
- The squad leader records and issues the fire command to the squad.
- The squad repeats the fire command.
- The ammunition bearer prepares the round according to the fire command.
- The squad leader inspects the round before it is passed to the assistant gunner. The ammunition bearer holds the round with both hands (palms up) near each end of the round body (not on the fuze or the charges).

Note. The mortar squad must communicate effectively to reduce the probability of human error when selecting the appropriate round type, number of charges, and fuze setting. Whenever possible, the squad leader physically inspects each round before firing. If the situation does not allow for physical inspection, the squad leader must visually inspect the round and verbally confirm and acknowledge with the assistant gunner that the round or rounds are all correct. The gunner remains the last control measure to verify that all data for the fire mission is without error (deflection and elevation, sight picture, and round).

- The assistant gunner checks the round for correct charges, fuze tightness, and fuze setting.
- When both the gun and the round(s) have been determined safe and ready to fire, the squad leader gives the following command to the FDC, "NUMBER (number of mortar) GUN, UP."
- The ammunition bearer holds the round with the fuze pointed to his left. By pivoting his body to the left, the assistant gunner accepts the round from the ammunition bearer with the right hand under the round and left hand on top of the round.
- Once the assistant gunner has the round, he keeps two hands on it until it is fired.

CAUTION
The assistant gunner is the only member of the mortar squad who loads and fires the round.

Chapter 1

- The squad leader commands, "HANG IT, FIRE" according to the method of fire given by the FDC.

Note. Until the baseplate is seated firmly, remove the sight from the mortar before firing each round. Standing on the baseplate is prohibited.

- The assistant gunner holds the round in front of the muzzle at about the same angle as the cannon. The gunner conducts final verification that the appropriate charge is set. (On rounds that have been previously handled and repackaged, the potential for charges to fall off exist, especially when being handled in the limited confines of a vehicle.) At the command, "HANG IT," the assistant gunner guides the round into the cannon (tail end first) to a point beyond the narrow portion of the body (about three-quarters of the round) being careful not to hit the primer or charges or disturb the lay of the mortar.
- Once the round is inserted into the cannon the proper distance, the assistant gunner shouts, "NUMBER (number of mortar) GUN, HANGING."
- At the command, "FIRE," the assistant gunner releases the round by pulling both hands down and away from the outside of the cannon. The assistant gunner ensures that he does not take his hands across the muzzle of the cannon while dropping the round.
- Once the round is released, the gunner and assistant gunner pivot their bodies so that they are both facing away from the blast.
- The assistant gunner pivots to his left and down toward the ammunition bearer, ready to accept the next round to be fired (unless major movements of the bipod require him to lift and clear the bipod off the ground).
- Subsequent rounds are fired based on the FDC fire commands.
- The assistant gunner ensures the round has fired safely before attempting to load the next round.
- The assistant gunner does not shove or push the round down the cannon. The round slides down the cannon under its own weight, strikes the firing pin, ignites, and fires.
- The assistant gunner and gunner, as well as the remainder of the mortar squad, keep their upper body below the muzzle until the round fires to avoid muzzle blast.
- During a fire for effect (FFE), the gunner tries to level all bubbles between each round, ensuring his upper body is away from the mortar and below the muzzle when the assistant gunner announces, "HANGING," for each round fired.
- The assistant gunner informs the squad leader when all rounds for the fire mission are expended, and the squad leader informs the FDC when all of the rounds are completed; for example, "NUMBER TWO GUN, ALL ROUNDS COMPLETE."
- When permissible or as determined by the FDC, the charge and elevation may be manually 'bumped' to the highest allowable data for the given target criteria. This method facilitates a rapid and solid settling of the baseplate and conserves ammunition. With an increased elevation and a higher charge, the recoil of the weapon system may offer a more uniform settle of the baseplate.

CAUTION

Wearing gloves can hinder hanging of the cartridges and is SOP-dependent according to local weather policies.

Introduction

> **WARNING**
>
> Keep feet clear of the mortar baseplate, as a rapidly seating baseplate can cause serious injury.

Note. Until the baseplate is seated firmly, remove the sight from the mortar before firing each round.

Target Engagement

1-65. Target engagement is achieved through fire commands, which are the technical instructions issued to mortar squads. The basis for these commands is the data processed in the FDC. There are two types of commands: initial fire commands, issued to start a fire mission; and subsequent fire commands, issued to change firing data and to cease firing. The elements of both commands follow the same sequence. However, subsequent commands include only elements that are changed, except for the elevation element, which always is announced.

1-66. A correct fire command is brief and clear, and includes all the elements necessary for accomplishing the mission. The commands are sent to the section leader or individual squad leaders if the section is dispersed, by the best available means. To limit errors in transmission, the person receiving the commands at the mortar position repeats each element as it is received. The sequence for the transmission of fire commands is shown in table 1-3.

Table 1-3. Sequence for transmittal of fire commands

Sequence	Example
Mortars to follow	Section
Shell and fuze	high explosive (HE) quick
Mortars to fire	Number two
Method of fire	One round
Deflection	Two eight zero zero
Charge	Charge 3
Time on target (TOT) fuze	28.4
Elevation	One two seven zero
When Ready	final protective fire (FPF)

Note. All fire commands will follow this sequence. Elements not necessary for the proper conduct of fire are omitted.

EXECUTION OF FIRE COMMANDS

1-67. The various fire commands are explained in the following paragraphs.

1-68. The **mortars to follow** element serves two purposes: it alerts the section for a fire mission and designates the mortars that are to follow the commands. The command for all mortars in the section to follow the fire command is "SECTION." Commands for individual or pairs of mortars are given a number (ONE, TWO, and so forth).

1-69. The **shell and fuze** element alerts the ammunition bearers (or assistant gunner for the 81-mm and 120-mm mortars without an available ammunition bearer) as to what type of ammunition and fuze action to prepare for firing ("HE QUICK, HE DELAY, HE PROXIMITY, and so forth").

1-70. The **mortar(s) to fire** element designates the specific mortar(s) to fire. If the mortars to fire are the same as the mortars to follow, this element is omitted. The command to fire an individual mortar or any combination of mortars is "NUMBER(s) ONE, (THREE, and so forth)."

1-71. In the **method of fire** element, the mortar(s) designated to fire in the preceding element is told how many rounds to fire, how to engage the target, and any special control desired. The number and type ammunition to be used in the fire-for-effect phase are included. Examples of method of fire are volley fire, section right (left), traversing and searching fire, "at my command," deflection and charge, time, elevation and when ready.

Volley Fire

1-72. A volley can be fired by one or more mortars. The command for volley fire is "(so many) ROUNDS." Once all mortars are reported up, they fire on the platoon sergeant's (section leader's) or squad leader's command. If more than one round is being fired by each mortar, the squads fire the first round on command and the remaining as rapidly as possible consistent with accuracy and safety, and without regard to other mortars. If a specific time interval is desired, the command is "(so many) ROUNDS AT (so many) SECONDS INTERVAL," or "(so many) ROUNDS PER MINUTE." In this case, a single round for each mortar, at the time interval indicated, is fired at the platoon sergeant's (section leader's) or squad leader's command.

Section Right (Left)

1-73. This is a method of fire in which mortars are discharged from the right (left) one after the other, normally at 10-second intervals. The command for section fire from the right (left) flank at intervals of 10 seconds is "SECTION RIGHT (LEFT), ONE ROUND." Once all mortars are reported up, the platoon sergeant (section leader) gives the command, "FIRE." For example, "SECTION RIGHT, ONE ROUND." The platoon sergeant (section leader) commands, "FIRE ONE;" then 10 seconds later, "FIRE TWO," and so forth.

1-74. If the section is firing a section left, the fire begins with the left-most gun and works to the right. The command "LEFT (RIGHT)" designates the flank from which the fire begins. The platoon sergeant (section leader) fires a section right (left) at 10-second intervals unless told differently by the FDC; for example, "SECTION LEFT, ONE ROUND, TWENTY-SECOND INTERVALS."

1-75. When continuously firing at a target is desired, the command is, "CONTINUOUS FIRE." When maintaining a smoke screen is desired, firing a series of sections right (left) may be necessary. In this case, the command is, "CONTINUOUS FIRE FROM THE RIGHT (LEFT)."

The platoon sergeant (section leader) then fires the designated mortars consecutively at 10-second intervals unless a different time interval is specified in the command.

1-76. Changes in firing data (deflections and elevations) are applied to the mortars in turns of traverse or elevation so as not to stop or break the continuity of fire; for example, "NUMBER ONE, RIGHT THREE TURNS. NUMBER TWO, UP ONE TURN." When continuous fire is given in the fire command, the platoon sergeant (section leader) continues to fire the section until the FDC changes the method of fire or until the command, "END OF MISSION" is given.

Traversing and Searching Fire

1-77. In **traversing fire**, rounds are fired with a designated number of turns of traverse between each round. The command for traversing fire is, "(so many) ROUNDS, TRAVERSE RIGHT (LEFT) (so many) TURNS." At the platoon sergeant's (section leader's) or squad leader's command, "FIRE," all mortars fire one round, traverse the specified number of turns, fire another round, and continue this procedure until the number of rounds specified in the command have been fired.

1-78. **Searching fire** is fired the same as volley fire except that each round normally has a different range. No specific order is followed in firing the rounds; for example, the assistant gunner does not start at the shortest range and progress to the highest charge or vice versa, unless instructed to do so. Firing the rounds in a definite sequence (high to low or low to high) establishes a pattern of fire that can be detected by the enemy.

At My Command and Do Not Fire

1-79. If the FDC wants to control the fire, the command "AT MY COMMAND" is placed in the method of fire element of the fire command. Once all mortars are reported up, the platoon sergeant (section leader) reports to the FDC, "SECTION READY." The FDC then gives the command, "FIRE."

1-80. The FDC can command "DO NOT FIRE" immediately following the method of fire. "DO NOT FIRE" then becomes a part of the method of fire. This command is repeated by the platoon sergeant (section leader). As soon as the weapons are laid, the platoon sergeant (section leader) reports to FDC that the section is laid. The command for the section of fire is the command for a new method of fire that is not followed by "DO NOT FIRE."

Deflection and Charge

1-81. The **deflection** element gives the exact deflection setting to be placed on the mortar sight. It is announced in four digits, and the word "DEFLECTION" always precedes the sight setting; for example, "DEFLECTION, TWO EIGHT FOUR SEVEN". When the mortars are to be fired with different deflections, the number of the mortar is given and then the deflection for that mortar; for example, "NUMBER THREE, DEFLECTION TWO FOUR ZERO ONE."

1-82. The **charge** element gives the charge consistent with elevation and range as, determined from the firing tables; for example, "CHARGE FOUR." The word CHARGE always precedes the amount; for example, "ONE ROUND, CHARGE FOUR."

Time, Elevation, and When Ready

1-83. The computer tells the ammunition bearer the exact time setting to place on the proximity (PRX) fuze; mechanical time, superquick (MTSQ); and mechanical time (MT) fuze. The command for time setting is "TIME (so much);" for example, "TIME TWO SEVEN." The command for a change in time setting is a new command for time.

1-84. The **elevation** element is given in the fire command. It serves two purposes: it gives the exact elevation setting to place on the mortar sight, and it gives permission to fire when the method of fire is the when ready method.

1-85. **When ready**, the squad leader verifies the mortar is layed on correct data and may commence firing under all mission types except for "AT MY COMMAND" and "DO NOT FIRE."

Note. The digital system displays "WHEN READY" automatically upon mission data transmission.

Arm-and-Hand Signals

1-86. When giving the commands "FIRE" or "CEASE FIRING," the section leader or squad leader uses both arm-and-hand signals and voice commands (see figure 1-6).

1-87. The signal for "I AM READY" or "ARE YOU READY?" is to extend the arm toward the person being signaled. Then, the arm is raised slightly above the horizontal, palm outward.

1-88. The signal to start fire is to drop the right arm sharply from a vertical position to the side. When the section leader wishes to fire a single mortar, he points with his arm extended at the mortar to be fired, then drops his arm sharply to his side.

1-89. The signal for cease firing is to raise the hand in front of the forehead, palm to the front, and to move the hand and forearm up and down several times in front of the face.

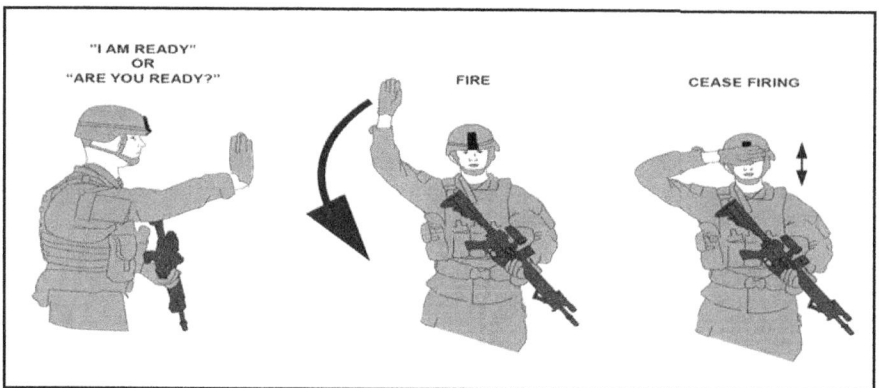

Figure 1-6. Arm-and-hand signals for ready, fire, and cease-fire

Subsequent Fire Commands

1-90. Only the elements that change from the previous fire command are announced in subsequent fire commands except for the elevation element (command to fire), which is always announced in every fire command.

Deflection and Elevation

1-91. Changes in direction are given in total deflection to be placed on the sight; for example, "DEFLECTION TWO EIGHT ONE TWO."

1-92. When a change is made in mortars to fire or in the method of fire, the subsequent command includes one or both of these elements, and the elevation. When the elevation does not change, the command "ELEVATION (so many mils)" is given (same as that given in the previous command).

Cease Firing/Check Fire and End of Mission

1-93. To interrupt firing, the "CEASE FIRING" or "CHECK FIRE," command is given. "CEASE FIRING" indicates to the section the completion of a fire mission, but not necessarily the end of the alert. Firing is renewed by issuing a new initial fire command. "CHECK FIRE" indicates a temporary cessation of firing and allows firing to be resumed with the same data by the command "RESUME FIRING" or by a subsequent fire command.

1-94. So that the mortar squads can relax between fire missions, the end of the alert is announced by the command, "END OF MISSION." All gunners then lay their mortars, as directed by the FDC. Upon completion of a fire mission, all mortars normally lay on final protective fire (FPF) data, unless otherwise directed. The platoon sergeant (section leader) is responsible for ensuring that the mortars are laid on FPF data and that the prescribed amount of ammunition for the FPF is prepared and on position.

REPETITION AND CORRECTION OF FIRE COMMANDS

1-95. If the platoon sergeant (section leader) or squad member fails to understand any elements of the fire command, he can request that element be repeated; for example by stating, "SAY AGAIN, DEFLECTION," "ELEVATION," and so forth. To avoid further misunderstanding, the repeated element is prefaced with "I SAY AGAIN, DEFLECTION (repeats mils)."

Initial and Subsequent Fire Commands

1-96. In an initial fire command, an incorrect element is corrected by stating, "CORRECTION," and giving only the corrected element.

1-97. In a subsequent command, an incorrect element is corrected by stating, "CORRECTION," and then repeating all of the subsequent commands. (The term correction cancels the entire command.)

ERROR REPORTS

1-98. When any squad member discovers that an error has been made in firing, the squad leader is immediately notified, who then notifies the FDC. Such errors include, but are not limited to, incorrect deflection or elevation settings, incorrect laying of the mortar, or ammunition improperly prepared for firing.

1-99. Misfires are also reported this way. Errors should be promptly reported to the FDC to prevent loss of time in determining the cause and required corrective action.

Mortar Squad Communication

1-100. Internal communication among the squad or squad members is imperative. Due to the nature of an indirect fire Infantryman's (military occupation specialty 11Cs) mission, squads must be fast, accurate, and above all else—safe. Whenever possible, the squad leader should physically inspect every round prior to being fired. However, this is not realistic under normal circumstances, especially when the squad leader is acting as the FDC or in a confined space such as the Stryker mortar vehicle or an M1064 mortar carrier. During hours of limited visibility or darkness and particularly during an immediate suppression or fire for effect, the ability for the squad leader to inspect each round prior to firing is further complicated and may compromise the intended or required effects of the fire mission, primarily in terms of speed.

1-101. Fire commands from the FDC (squad leader) to the squad is echoed, but under arduous conditions or duress, despite the commands being repeated by the squad, squad members run the risk of human error. The given deflection and elevation data can be inaccurately indexed on the sight unit; the wrong charge or fuze can be set on a mortar round, even the wrong type of round may be selected, or number of rounds intended for the fire mission. Cross-talk for confirmation and acknowledgement significantly reduces these probabilities.

1-102. Whenever possible the squad leader should inspect both the mortar rounds and the lay of the gun (data indexed for deflection and elevation, leveled bubbles, and proper sight picture). Under circumstances where this may not be feasible, mortar squads must be trained to ensure a system of checks and balances.

1-103. Ammunition bearers must verbally call-off the type of round, the number of charges, and the appropriate fuze setting AFTER preparing the round(s). The assistant gunner inspects the round, and verbally acknowledges its correctness, ensuring the gunner and the squad leader has his attention. The round or rounds are only safe to be fired once the squad leader confirms that all data is correct. The gunner is the final safety check prior to the round being 'hung'

Note. The probability of a propelling charge falling off the round during a fire mission is a real possibility between the round data being checked and secured for firing, especially for mortar rounds that have previously been handled and repackaged.

1-104. This common condition heightens the necessity for communication among squad members. By recording the data being given to them by the FDC (and through the squad leader), mortar squads further ensure that the data was not misinterpreted reducing the need for asking commands to be repeated several times, which may only add confusion. Visual inspections and verbal confirmation and acknowledgement tremendously lessen the inherent risks associated with actions being performed by the mortar squad.

Night Fire

1-105. When firing the mortar at night, the mission dictates whether noise and light discipline are to be sacrificed for speed. To counteract the loss of speed for night firing, the gunner must consider presetting both fuze and charge for ILLUM rounds with the presetting of charges for other rounds. The procedure for manipulating the mortar at night is the same as during daylight operations. To assist the gunner in these manipulations, the sight reticle is illuminated and the aiming posts are provided with lights.

1-106. The instrument lights illuminate the reticle of the sights and make the vertical cross hairs visible. The scales are illuminated so that the deflection and elevation are readable.

Introduction

1-107. An aiming post light is placed on each aiming post to enable the gunner to see the aiming posts. Aiming posts are placed out at night similar to the daylight procedure. Attach the lights to the posts before they can be seen and positioned by the gunner. The gunner must issue commands such as, "NUMBER ONE, MOVE RIGHT, LEFT, HOLD, DRIVE IN, POST CORRECT." Tilt in the posts is corrected at daybreak. Some of the distance to the far post can be sacrificed if it cannot be easily seen at 100 meters. However, the near post should be positioned about half the distance to the far post from the mortar. The far post light should be a different color from the one on the near post and be positioned so it appears slightly higher. Adjacent squads should alternate post lights to avoid laying on the wrong posts; for example, "FIRST SQUAD, NEAR POST—GREEN LIGHT, FAR POST—ORANGE LIGHT; SECOND SQUAD, NEAR POST—ORANGE LIGHT, FAR POST—GREEN LIGHT." (The M58 light is green and the M59 light is orange.)

1-108. The mortar is laid for deflection by placing the vertical cross hair of the sight in the correct relation to the center of the lights attached to the aiming posts. The procedure for laying the mortar is the same as discussed earlier this chapter.

1-109. The night lights are used to align the aiming posts without using voice commands. This requires the following actions:

- The gunner directs the ammunition bearer to place out the aiming posts. The ammunition bearer moves out 100 meters and turns on the night light of the far aiming post. The gunner holds a night light in his right (seen as left) hand and by moving the light to the right (seen as left), directs the ammunition bearer to move to the right (seen as left). To ensure that the ammunition bearer sees the light moving only in the desired direction, the gunner places his thumb over the light when returning it to the starting position. The gunner continues to direct the ammunition bearer to move the aiming post until it is aligned properly.
- The gunner moves the light a shorter distance from the starting position when he wants the ammunition bearer to move the aiming post a short distance.
- The gunner holds the light over his head (starting position) and moves the light to waist level when he wants the ammunition bearer to place the aiming post into the ground. In returning the light to the starting position, the gunner covers the light with his thumb to ensure that the ammunition bearer sees the light move only in the desired direction.
- The gunner uses the same procedure described above when he wants the ammunition bearer to move the aiming post light to a position corresponding to the vertical hairline in the sight after the aiming post has been placed into the ground.
- The gunner reverses the procedure described above when he wants the ammunition bearer to take the aiming post out of the ground. The gunner places the uncovered light at waist level and moves it to a position directly above his head. He then directs alignment as required.
- When the gunner is satisfied with the alignment of the aiming posts, he signals the ammunition bearer to return to the mortar positions by making a circular motion with the instrument light.

Note. When the night light is used to signal, the gunner directs the light toward the ammunition bearer. In hours of darkness, great care must be taken to ensure that all firing data is correct and according to the fire mission at hand. The use of a red-lens flashlight to inspect mortar rounds, and the cross-talk among squad members and squad leader is crucial to the safe conduct of night missions

This page intentionally left blank.

Chapter 2

Mortar Equipment and Employment

Proper employment of sighting and fire control equipment ensures effective fire against the enemy. This chapter describes this equipment and its applications.

BORESIGHTING

2-1. Boresighting is a vital measure to enhance accuracy that must be conducted. Procedures include manual and digital boresighting, depending on the equipment assigned. Boresights are adjusted by the manufacturer and should not require readjustment as a result of normal field handling. (Refer to the most updated technical manual [TM].)

PRINCIPLES OF OPERATION

2-2. Except for the M115, the boresight is constructed so that the telescope line of sight lies in the plane established by the center lines of the V-slides. When properly secured to a mortar barrel, the centerline of the contacting V-slide is parallel to the centerline of the barrel. Further, when the cross-level vial is centered it indicates that the center lines of both slides—the elbow telescope and the barrel lie in the same vertical plane. Therefore, the line of sight of the telescope coincides with the axis of the barrel, regardless of which V-slide of the boresight is contacting the barrel. The elevation vial is constructed with a fixed elevation of 800 mils.

Note. The base gun gunner should boresight all mortar systems in the section. To eliminate variations in sight alignment and provide consistency in boresighting.

TABULATED DATA

2-3. The tabulated data of the M45-series and M115-series boresights are shown in table 2-1.

Table 2-1. Tabulated boresight data for M45- and M115-series

M45-Series		M115-Series	
Weight	2.5 pounds	Weight	5 ounces
Field of view	12 degrees	Field of view	17 degrees
Magnification	3 power	Magnification	1.5 power

M45-Series Boresight

2-4. The M45-series boresight (see figure 2-1) detects deflection and elevation errors in the sight and consists of an elbow telescope, telescope clamp, body, two strap assemblies, and clamp assembly.

2-5. The elbow telescope establishes a definite line of sight while the telescope clamp maintains the line of sight in the plane established by the centerline of the V-slides.

2-6. The body incorporates two perpendicular V-slides. It contains level vials (preset at 800 mils elevation) that are used to determine the angle of elevation of 800 mils and if the V-slides are in perpendicular positions. It also provides the hardware to which the straps are attached.

2-7. Two strap assemblies are supplied with each boresight and are marked for cutting in the field to the size required for any mortar. The clamp assembly applies tension to the strap assemblies to secure the boresight against the mortar barrel.

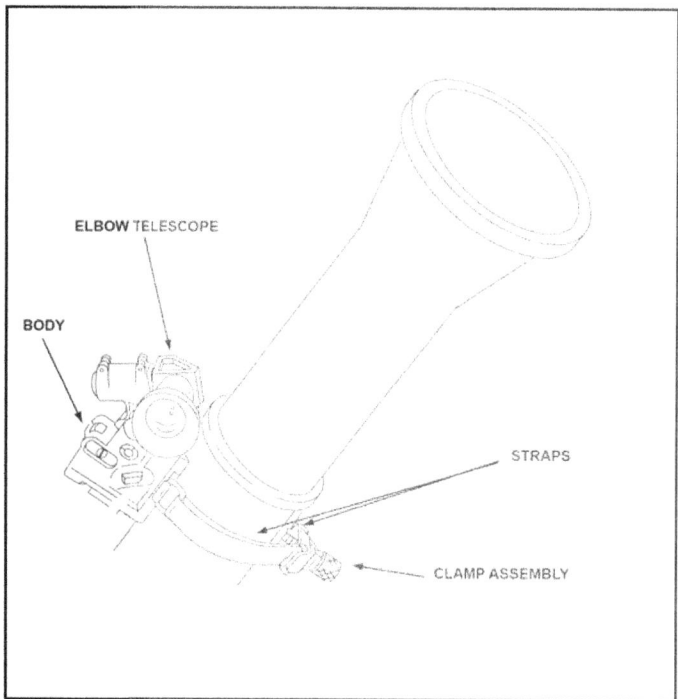

Figure 2-1. M45 boresight

Installation

2-8. Installation procedures for the M45 boresight are as follows:
- Remove the boresight, clamp assembly, and straps from the carrying case. Grasp the boresight by the body to prevent damaging the telescope.
- Place the ring over the hook and attach the strap snap to the eye provided on the strap shaft.
- If necessary, release the catches and reset the straps to the proper length.

Mortar Equipment and Employment

- Remove any burrs or projecting imperfections from the seating area of the mortar barrel to ensure proper seating of the boresight. Attach the boresight to the barrel below and touching the upper stop band on the M252/M252A1 mortar.

M115 BORESIGHT

2-9. The M115 boresight (see figure 2-2) detects deflection and elevation errors in the sight. The boresight has three plungers that keep it in place when mounted in the muzzle of the barrel and the telescope has a field of view of 17 degrees (302 mils) with a magnification of 1.5 unity power. The elevation bubble levels are only at 800 mils.

Note. The M115 boresight is only used with 60-mm mortar systems.

2-10. The components of the M115 boresight are the body, telescope, and leveling bubbles (one for cross-leveling and one for elevation).

2-11. A second cross-level bubble is used as a self-check of the M115. After leveling and cross-leveling, the M115 can be rotated 180 degrees in the muzzle until the second cross-level bubble is centered. The image of the boresight target should not vary in deflection. A large deviation indicates misalignment between the cross-level bubble and lenses, and the boresight should be turned into maintenance.

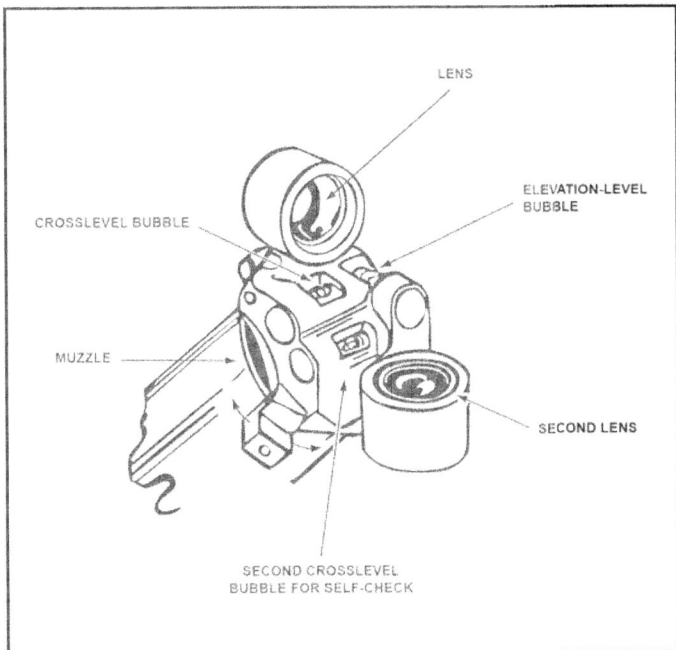

Figure 2-2. M115 boresight

Chapter 2

SIGHT CALIBRATION

2-12. Always calibrate the mortar sight to the mortar on which it will be mounted. This procedure is necessary since the sight socket that receives the sight unit is a machined part and varies in accuracy with each mortar.

2-13. There is no set rule for frequency of calibration. The sight\t 2-13 should be calibrated each time the mortar is mounted in a new location because movement might disturb the setting of the elevation and deflection scales. Time available and accuracy dictate the frequency of calibration.

BORESIGHT METHOD OF CALIBRATION

2-14. Once the mortar has been mounted, place the M67 sight unit into position in the sight socket. Place a deflection of 3200 mils and an elevation of 0800 mils on the scales; for the M252/M252A1 mortar, place a deflection of 0 mils on the sight.

2-15. The **distant aiming point method** is the preferred and most accurate technique when boresighting. The gunner identifies an aiming point at least 200 meters from the gun position and follows the procedures for proper boresighting.

2-16. Align the vertical crosshairs of the sight on an aiming point, look through elbow telescope and pick out a distant aiming point (DAP). If necessary, move the mortar to place the crosshairs on the left vertical edge of the DAP. Do not use the traversing mechanism, but physically pick up the mortar to align it on the aiming point. Make a visual check of the mortar for cant; if cant exists, remove the cant and lay again, if necessary.

2-17. The **25-meter sight box method** should only be used if the situation does not allow for the gunner to identify a DAP. Using the sight box as an aiming point the gunner uses arm-and hand signals to place the sight box in alignment with the vertical crosshairs of the boresight.

2-18. There are two circles on the M67 sight case: the small circle (six inches) is used with the legacy 60-mm/legacy 81-mm systems; the large circle (8.28-inches with writing) is used with the M224A1 60-mm and M252A1 81-mm systems. The 120-mm mortar does not use the circles but uses the outer edges of the sight case for this method.

Elevation Setting

2-19. Use the following procedures to set the elevation:
- Install the boresight on the mortar barrel. Center the cross-level vial by rotating the boresight slightly on the outside diameter of the mortar barrel. Slight movements are made by loosening the clamp screw and lightly tapping the boresight body. When the bubble centers, tighten the clamp screw.
- Elevate or depress the mortar barrel until the boresight elevation level vial is centered. The mortar is now set at 800 mils (45 degrees) elevation.
- Using the elevation micrometer knob, elevate or lower the sight unit until the elevation level bubble is centered. If necessary, cross-level the sight unit.
- Recheck all level bubbles.

Mortar Equipment and Employment

- The reading on the coarse elevation scale of the sight unit should be 800 mils and the reading on the elevation micrometer scale should be 0. If adjustment is necessary, proceed as indicated below:

   Loosen the two screws on the elevation micrometer knob and slip the elevation micrometer scale until the 0 mark on the micrometer scale coincides with the reference mark on the housing. Tighten the two screws to secure the micrometer scale.

Note. Do not adjust the M64-/M67-series sight unit coarse elevation scale. If it does not line up with the 0800-mil mark, turn it in to maintenance.

- Recheck all level bubbles.

Deflection Setting

2-20. Use the following procedures to set the deflection:

- Check again to ensure that the sight setting reads 3200 on the fixed deflection scale and elevation 800 mils. Set zero deflection for the M252/M252A1 mortar.

- Traverse the mortar no more than two turns of center of traverse (four turns for the 120-mm mortar) and align the vertical crosshairs of the boresight on the original aiming point. Adjust the boresight to keep the cross-level bubble centered since the mortar could cant during traversing. (If the mortar initially is mounted on the aiming point, it decreases the traverse needed to align the cross hair on the aiming point.) Also, the elevation level bubble may need to be leveled.

- After the boresight is aligned on the aiming point, level the sight by centering the cross-level bubble. Rotate the deflection micrometer knob until the sight is aligned on the aiming point. The coarse deflection scale should read 3200 mils and the micrometer scales should read 0. If adjustment is necessary, loosen the two screws on the deflection micrometer knob and slip the micrometer deflection scale until the arrow on the index is aligned with the zero mark on the micrometer scale.

- To verify proper alignment, remove the boresight and place it in position on the underside of the cannon. Center the boresight cross-level bubble and check the vertical cross hair to see if it is still on the aiming point. If cant exists, the vertical cross hair of the boresight is not on the aiming point. This indicates that the true axis of the bore lies halfway between the aiming point and where the boresight now is pointing.

- To correct this error, look through the boresight and traverse the mortar onto the aiming point. If bubbles are level, use the deflection micrometer knob and place the vertical cross hair of the sight back onto the aiming point. With the sight in this position, index half of the mil variation between the sight and boresight. Slip zero on the micrometer scale to the index mark: for example, the mil variation is 10 mils and one-half of this value is five mils. Loosen the two screws on the deflection micrometer and index zero.

- Check all level bubbles, sight unit, and boresight.

- With a deflection on the micrometer scale of half the value of the original mil variation, both the sight unit and boresight are on the aiming point. If an error exists, repeat the procedure outlined above.

Removal

2-21. Use the following procedures to remove the boresight:

- Loosen the clamp screw, releasing the boresight from the barrel.

- Swing the elbow telescope until it is about parallel with the elevation level bubble.

Chapter 2

- Release the clamp assembly and straps by removing the ring from the hook and strap shaft.
- Stow the clamp assembly and straps in the corner compartment. Put the boresight in the center compartment of the carrying case.

Boresight Mounted on a Mortar Carrier Vehicle

2-22. For the MFCS software to properly compute the boresight values, the mortar trunnions must be leveled. In a tactical environment, this is most easily accomplished using a gunner's quadrant. To make the process easier to perform, the vehicle should be parked on a slight incline, roughly 20 to 50 mils, with the front of the Stryker mortar vehicle higher than the rear.

2-23. The incline may be measured with the gunner's quadrant placed on top of the mortar's traversing mechanism gearbox, or traversing gear ring, and aligned with the fore and aft axis of the carrier. The line of fire indicated on the gunner's quadrant (see figure 2-3) should point toward the front of the vehicle.

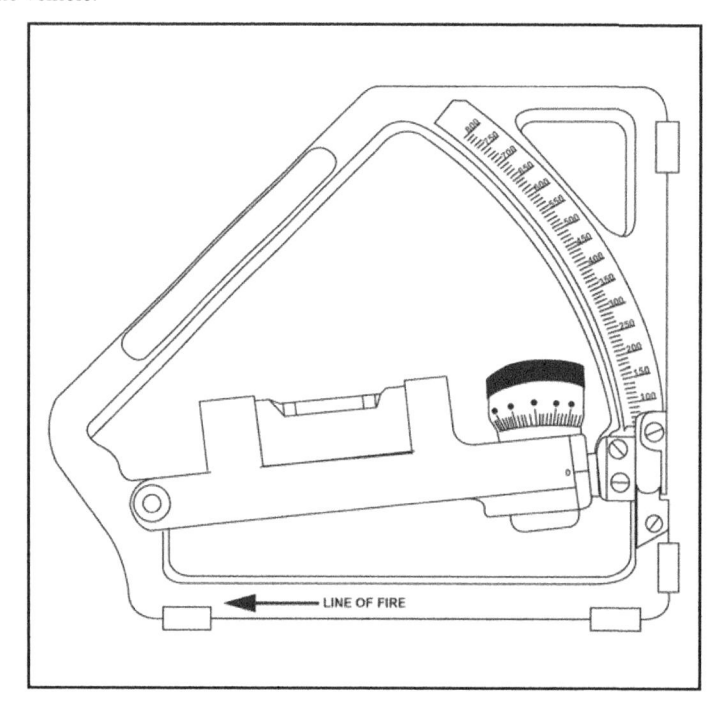

Figure 2-3. Gunner's quadrant sight

Note. When using the gunner's quadrant, be sure the mating surfaces are free of any burrs, projections, or paint that would prevent the quadrant from seating properly. The gunner's quadrant must be held against the three registration pins on the seating pads to achieve an accurate reading.

M2 COMPASS

2-24. The M2 compass (see figure 2-4) is a multipurpose instrument used primarily to obtain azimuths and angles of sight. It measures grid azimuths after the instrument has been declinated for the local area.

Note. Refer to TM 9-1290-333-15 for more information.

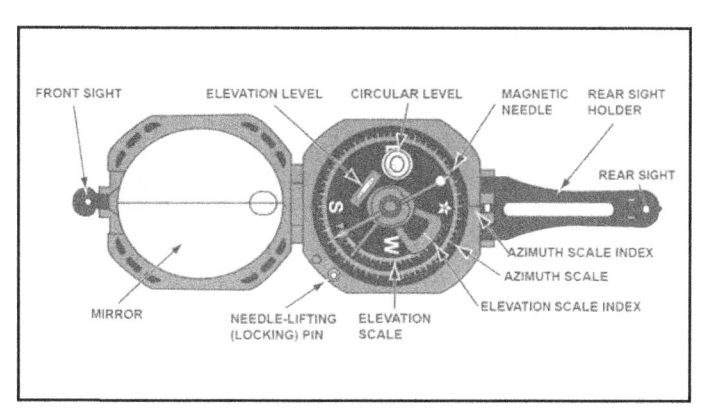

Figure 2-4. M2 compass (top view)

CHARACTERISTICS AND COMPONENTS

2-25. The main characteristics of the M2 compass include the following:
- Angle-of-sight scale: 1200-0-1200 mils.
- Azimuth scale: 0 to 6400 mils.
- Dimensions closed: 70-mm by 28.5-mm (2 3/4 by 1 1/8 inches).
- Weight: Eight ounces.

2-26. The principal parts of the compass include the following:
- Compass body assembly.
- Angle-of-sight mechanism.
- Magnetic needle and lifting mechanism.
- Azimuth scale and adjuster.
- Front and rear sight.

2-27. The **compass body assembly** consists of a nonmagnetic body and a circular glass window that covers the instrument and keeps dust and moisture from its interior, protecting the compass needle and angle-of-sight mechanism. A hinge assembly holds the compass cover in the position in which it is placed. A hole in the cover coincides with a small oval window in the mirror on the inside of the cover. A sighting line is etched across the face of the mirror.

2-28. The **angle-of-sight mechanism** is attached to the bottom of the compass body. It consists of an actuating (leveling) lever located on the back of the compass, a leveling assembly with a tubular elevation level, and a circular level. The instrument is leveled with the circular level to read

Chapter 2

azimuths and with the elevation level to read angles of sight. The elevation (angle-of-sight) scale and the four points of the compass, represented by three letters and a star, are engraved on the inside bottom of the compass body. The elevation scale is graduated in two directions; in each direction, it is graduated from 0 to 1200 mils in 20-mil increments and numbered every 200 mils.

2-29. The **magnetic needle assembly** consists of a magnetized needle and a jewel housing that serves as a pivot. The north-seeking end of the needle is white. (The newer compasses have the north and south ends of the needle marked "N" and "S" in raised, white lettering.) On some compasses, a thin piece of copper wire is wrapped around the needle for counterbalance. A lifting pin projects above the top rim of the compass body. The lower end of the pin engages the needle-lifting lever. When the cover is closed, the magnetic needle is lifted automatically from its pivot and held firmly against the window of the compass.

2-30. The **azimuth scale** is a circular dial geared to the azimuth scale adjuster. This permits rotation of the azimuth scale about 900 mils in either direction. The azimuth index provides a means of orienting the azimuth scale at 0 or the declination constant of the locality. The azimuth scale is graduated from 0 to 6400 in 20-mil increments and numbered at 200-mil intervals.

2-31. The **front sight** is hinged to the compass cover. It can be folded across the compass body, and the cover closed. The rear sight is made in two parts—a rear sight and a holder. When the compass is not being used, the **rear sight** and holder are folded across the compass body and the cover is closed.

OPERATION OF M2 COMPASS

2-32. Hold the compass steady to obtain accurate readings. The use of a sitting or prone position, a rest for the hand or elbows, or a solid nonmetallic support helps eliminate unintentional movement of the instrument. When being used to measure azimuths, the compass must not be near metallic objects.

> *Note.* Failure to distance the compass from metallic objects or power sources (overhead and underground power lines) can alter compass readings and cause inaccurate information data.

2-33. To measure a magnetic azimuth—

- Zero the azimuth scale by turning the scale adjuster.
- Place the cover at an angle of about 45 degrees to the face of the compass so that the scale reflection is viewed in the mirror.
- Adjust the front and rear sights to the desired position. Sight the compass using any of these methods:
    - Raise the front sight and the extended rear sight assembly perpendicular to the face of the compass. (See figure 2-5 and figure 2-6.) Sight over the tips of the front and rear sights. If the object is above the line of sighting, fold the rear sight toward the eye, as needed. The instrument is aligned correctly when, with the level centered, the operator sees the tips of the sights and the center of the object at the same time.
    - Raise the rear sight almost perpendicular to the face of the compass. Sight on the object through the opening in the rear sight holder and through the window in the cover. Keep the compass level, and raise or lower the eye along the opening in the rear sight holder until the black center line of the window bisects the object and the opening in the rear sight holder.

■ Fold the rear sight holder out parallel with the face of the compass with the rear sight perpendicular to its holder. Sight through or over the rear sight and view the object through the window in the cover. If the object sighted is at a lower elevation than the compass, raise the rear sight holder as needed. The compass is sighted correctly when the compass is level and the operator sees the black center line of the window bisecting the rear sight and the object sighted.

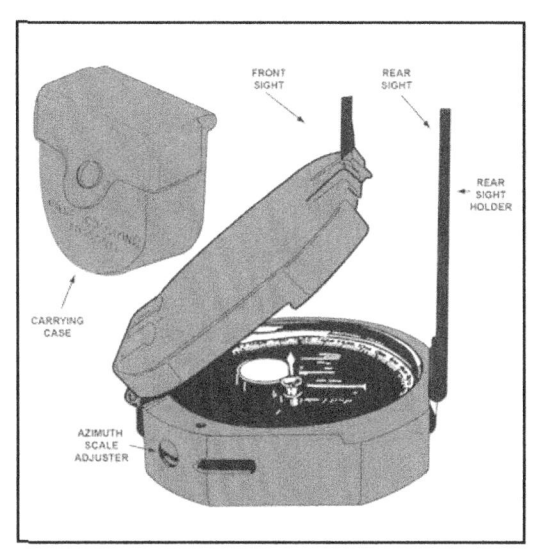

Figure 2-5. M2 compass (side view)

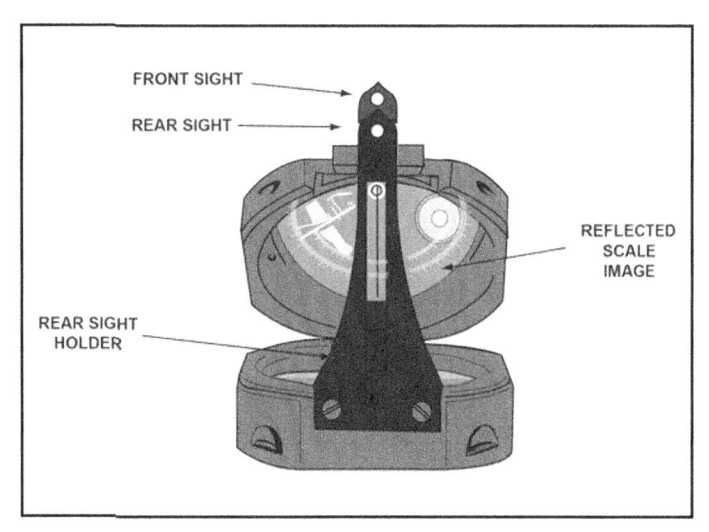

Figure 2-6. M2 compass (user's view)

● Hold the compass in both hands, at eye level, with the arms braced against the body and the rear sight near the eyes; for precise measurements, rest the compass on a nonmetallic stake or object as depicted in figure 2-6.

Chapter 2

- Level the instrument by viewing the circular level in the mirror and moving the compass until the bubble is centered. Sight on the object, look in the mirror, and read the azimuth indicated by the black (south) end of the magnetic needle.

2-34. To measure a grid azimuth—
- Index the known declination constant on the azimuth scale by turning the azimuth scale adjuster. Be sure to loosen the locking screw on the bottom of the compass. (The lightweight plastic M2 compass does not have a locking screw.)
- Measure the azimuth as described above. The azimuth measured is a grid azimuth.

2-35. To measure an angle-of-sight or vertical angle from the horizontal—
- Hold the compass with the left side down (cover to the left), and fold the rear sight holder out parallel to the face of the compass, with the rear sight perpendicular to the holder. Position the cover so that, when looking through the rear sight and the aperture in the cover, the elevation vial is reflected in the mirror.
- Sight on the point to be measured.
- Center the bubble in the elevation level vial (reflected in the mirror) with the level lever.
- Read the angle on the elevation scale opposite of the index mark. The section of the scale graduated counterclockwise from 0 to 1200 mils measures plus angles of sight. The section of the scale graduated clockwise from 0 to 1200 mils measures negative angles of sight.

2-36. To declinate the M2 compass from a surveyed declination station free from magnetic attractions—
- Set the M2 compass on an aiming circle tripod over the orienting station, and center the circular level.
- Sight in on the known, surveyed azimuth marker.
- Use the azimuth adjuster scale by rotating the azimuth scale until it indicates the same as the known, surveyed azimuth.
- Recheck the sight picture and azimuth to the known point. Once the sight picture is correct and the azimuth reading is the same as the surveyed data, the M2 is declinated.

2-37. To use the field-expedient method to declinate the M2 compass—
- Use the azimuth adjuster scale, set off the grid-magnetic (G-M) angle shown at the bottom of all military maps.
- Once the G-M angle has been set off on the azimuth scale, the M2 compass is declinated.

M2 AND M2A2 AIMING CIRCLES

2-38. Aiming circles, M2, and M2A2, are used to measure azimuth and elevation angles with respect to a preselected baseline. An aiming circle is a low-power telescope that is mounted on a composite body and contains a magnetic compass, adjusting mechanisms, and leveling screws for establishing a horizontal plane.

2-39. The instrument is supported by a baseplate for mounting on a tripod. Angular measurements in azimuth are indicated on graduated scales and associated micrometers. (Refer to TM 9-6675-262-10 for more information.)

SIGHT UNITS

2-40. The M67- and M64-series sight units are the standard sighting devices used with mortars. (Refer to TM 9-1010-223-10, TM 9-1015-249-10, or TM 9-1015-250-10 for more information.) The M64-series sight units still may be in use but are being replaced with the M67-series as they become unserviceable. The mortar sight unit (see figure 2-7) is the instrument in which the gunner sets a given deflection and elevation to engage targets. Lighting for night operations is provided by radioactive tritium gas contained in phosphor-coated glass vials within the sight unit.

2-41. The sight unit is lightweight and portable. It is attached to the bipod mount by means of a dovetail. Coarse elevation and deflection scales and fine elevation and deflection scales are used in conjunction with elevation and deflection knob assemblies to sight the mortar system. After the sight has been set for deflection and elevation, the mortar is elevated or depressed until the elevation bubble on the sight is level. The mortar then is traversed until a proper sight picture is seen (using the aiming posts as the aiming point) and the cross-level bubble is level. The mortar is laid for deflection and elevation when all bubbles are level within the outer red lines and the gunner has attained the proper sight picture. After the ammunition has been prepared, it is ready to be fired.

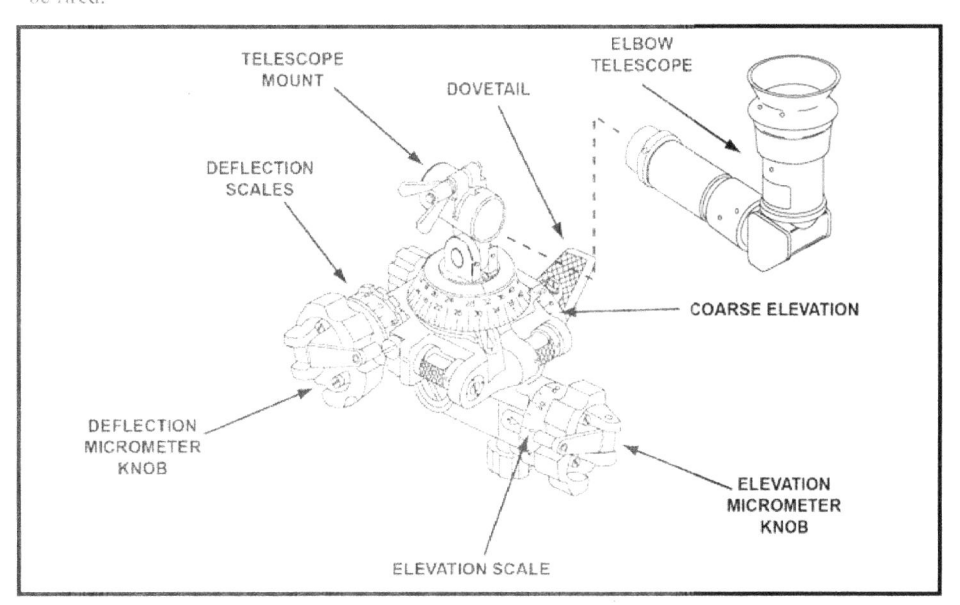

Figure 2-7. M67 sight unit

MAJOR COMPONENTS

2-42. The sight unit consists of two major components: the elbow telescope and the telescope mount. The M67 elbow telescope is 4.0-power, hermetically sealed with a tritium illuminated cross reticle.

2-43. The telescope mount, provided with tritium back-lighted level vials, indexes, and translucent plastic scales, is used to orient the elbow telescope in azimuth and elevation. The equipment data for the M67 sight unit are shown in table 2-2.

Chapter 2

CAUTION
When not in use, store the sight unit in its carrying case.

Table 2-2. M67 sight unit equipment data

Weight	2.9 pounds (1.3 kilograms)
Field of View	10 degrees
Magnification	4.0 X nominal, 3.5 effective
Length	5 3/8 inches (13.7 centimeters)
Width	4 3/8 inches (11.1 centimeters)
Height	8 1/2 inches (21.6 centimeters)
Light Source	Self-contained, radioactive tritium gas (H^3).

SIGHT SUBCOMPONENTS

2-44. The elbow telescope has an illuminated cross reticle. The sight mount has a dovetail, locking knobs, control knobs, scales, cranks, and locking latch.

2-45. The **dovetail** is compatible with standard U.S. mortars. When the dovetail is seated properly in the dovetail slot, the locking latch clicks. The locking latch is pushed toward the barrel to release the sight from the dovetail slot for removal.

2-46. The red locking **knobs** lock the deflection and elevation mechanisms of the sight during firing. The elevation and deflection micrometer knobs are large for easy handling. Each knob has a crank for large deflection and elevation changes.

2-47. All **scales** can be adjusted to any position. Micrometer scales are white. The elevation micrometer scale and fixed boresight references (red lines) above the coarse deflection scale and adjacent to the micrometer deflection scale are slipped by loosening slot-headed screws. Coarse deflection scales and micrometer deflection scales are slipped by depression and rotation. The coarse elevation scale is factory set and should not be adjusted at the Soldier level. The screws that maintain the coarse elevation scale are held in place with locking compound. If the screws are loosened and then tightened without reapplying the locking compound, the coarse elevation scale can shift during firing.

Note. If the index does not align with the coarse elevation scale within ±20 mils when boresighting at 800 mils, field-level maintenance should be notified.

2-48. Instrument lights are not needed when using the sight unit at night. Nine parts of the sight are illuminated by tritium gas:
- Telescope.
- Coarse elevation scale.
- Coarse elevation index arrow.
- Elevation vial.

- Fine elevation scale.
- Coarse deflection index arrow.
- Cross-leveling vial.
- Fine deflection scale.
- Coarse deflection scale.

> **CAUTION**
> When not in use, store the sight unit in its carrying case.

OPERATION OF SIGHT UNITS

2-49. The operation of all the U.S. sight units is similar. To attach the sight unit, insert the dovetail of the telescope mount into the sight socket. Press the locking lever inward, seat the mount firmly, and release the locking lever.

Note. Until the baseplate is seated firmly, remove the sight from the mortar before firing each round.

Setting the Deflection

2-50. Setting the correct deflection on the sight places the mortar, once the sight is back on the poles and leveled, in the direction commanded by the FDC.

2-51. To place a deflection setting on the sight, turn the deflection knob. Before attempting to place a deflection setting on the sight unit, ensure that the deflection locking knob is released. After placing a setting on the sight, lock the locking knob to lock the data onto the sight and to ensure that the scale does not slip during firing.

2-52. When setting the deflection, use the fixed coarse scale and the fixed micrometer scale to obtain the desired setting. Set the first two digits of the deflection using the coarse deflection scale and the last two using the deflection micrometer scale.

Note. The coarse scale and the micrometer scale are slip scales.

2-53. Setting a deflection on the deflection scale does not change the direction in which the barrel is pointing (the lay of the mortar). It only moves the vertical line off (to the left or right) the aiming poles. The deflection placed on the sight is the deflection announced in the fire command. Place a deflection on the sight before elevation.

Setting the Elevation

2-54. Setting the correct elevation on the sight places the mortar, once the sight is leveled, on the angle of fire commanded by the FDC. Setting an elevation on the elevation scale does not change the elevation of the mortar barrel.

2-55. To set for elevation, turn the elevation knob. This operates both the elevation micrometer and coarse elevation scales. Both scales must be set properly to obtain the desired elevation. To place an elevation of 1065 mils for example, turn the elevation knob until the fixed index opposite the moving coarse elevation scale is between the black 1000-mil and 1100-mil graduations on the scale (the graduations are numbered every 200 mils), and then use the micrometer knob to set the last two digits (65).

2-56. Before setting elevations on the sight, unlock the elevation locking knob. Once the elevation is placed on the sight, lock the elevation locking knob. This ensures the data placed on the sight does not change accidentally.

Correct Sight Picture and Laying for Direction

2-57. Sight the mortar on the aiming posts with the vertical reticle line aligned on the left side of the aiming post. Check to ensure the bubbles are level. Repeat adjustments until a correct sight picture is obtained with the bubbles level.

2-58. In laying the mortar for direction, the two aiming posts do not always appear as one when viewed through the sight(s). This separation is caused by either a large deflection shift of the barrel or a rearward displacement of the baseplate assembly caused by the shock of firing. (See figure 3-7 on page 3-10 for an example of a compensated sight picture.)

2-59. When the aiming posts appear separated, the gunner cannot use either one as the aiming point. To lay the mortar correctly, he establishes a compensated sight picture. To take a compensated sight picture, the gunner traverses the mortar until the sight picture appears with the left edge of the far aiming post placed exactly midway between the left edge of the near aiming post and the vertical line of the sight. This corrects for the displacement.

Note. A memory trick for correcting displacement is: Hey diddle, far pole in the middle.

2-60. The gunner determines if the displacement is caused by traversing the mortar or by displacement of the baseplate assembly. To do so, place the referred deflection on the sight and lay on the aiming posts. If both aiming posts appear as one, the separation is caused by traversing. Therefore, the gunner continues to lay the mortar as described, and does not realign the aiming posts. When the posts appear separated, the separation is caused by displacement of the baseplate assembly. The gunner notifies the squad leader, who requests permission from the section/platoon leader to realign the aiming posts.

Replace the Sight Unit in the Carrying Case

2-61. Before returning the sight unit to the carrying case, close the covers on the level vials and set an elevation of 1100 mils and deflection of 3800 mils on the scales.

2-62. Place the elbow telescope in the left horizontal position. All crank handles should be folded into the inoperative position.

CARE AND MAINTENANCE OF SIGHT UNITS

2-63. Although sight units are rugged, the units can become inaccurate or malfunction if abused or handled roughly. To properly care for and maintain the sight unit—

Mortar Equipment and Employment

- Avoid striking or otherwise damaging any part of the sight. Be particularly careful not to burr or dent the dovetail bracket. Avoid bumping the micrometer knobs, telescope adapter, and level vials. Keep the metal vial covers closed except when using the sight.
- Keep the sight in the carrying case when not in use. Keep it as dry as possible, and do not place it in the carrying case while it is damp.
- When the sight fails to function correctly, return it to the field-level maintenance unit for repair.

Note. Mortar squad members are not authorized to disassemble the sight.

- Keep the optical parts of the telescope clean and dry. Remove dust from the lens with a clean camel-hair brush. Use only lens cleaning tissue to wipe these parts. Do not use ordinary polishing liquids, pastes, or abrasives on optical parts. Use only authorized lens cleaning compound for removing grease or oil from the lens.
- Occasionally, oil only the sight locking devices by using a small quantity of light preservative lubricating oil. To prevent accumulation of dust and grit, wipe off excess lubricant that seeps from moving parts. Ensure that no oil gets on the deflection and elevation scales. (Oil removes the paint from the deflection scale.) No maintenance is authorized.

M67 Sight Unit Hazards

2-64. The M67 uses radioactive tritium gas (H^3) for illumination during night operations. The gas is sealed in glass tubes and is not hazardous when the glass tubes are intact. If there is no illumination, notify the chemical, biological, radiological, and nuclear officer, range management office (during peacetime), unit chain of command, and follow unit SOPs and local regulations for containment. (Refer to TC 4-02.1 for more information on first aid.)

> **WARNING**
>
> Do not try to repair or replace the radioactive material. If skin contact is made with tritium, wash the area immediately with nonabrasive soap and water.

2-65. Radioactive self-luminous sources are identified by warning labels (see figure 2-8, page 2-16), which should not be defaced or removed. If necessary, they must be replaced immediately.

Chapter 2

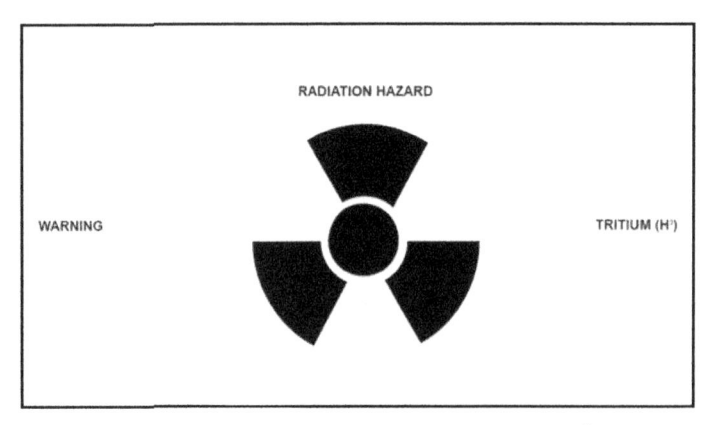

Figure 2-8. Warning label for tritium gas (H^3)

2-66. All radioactively illuminated instruments or modules that are defective must be evacuated to a sustainment-level maintenance activity. Defective items must be placed in a plastic bag and packed in the shipping container from which the replacement was taken. Spare equipment must be stored in the shipping container as received until installed on the weapon. Such items should be stored in an outdoor shed or unoccupied building.

OTHER EQUIPMENT

2-67. Other equipment required to operate and employ mortars are M14 and M1A2 aiming posts, M58 and M59 aiming post lights, and M16 and M19 plotting boards.

2-68. The Mortar Fire Control System (MFCS) or the lightweight handheld mortar ballistic computer (LHMBC) is also used to employ mortars.

M14 AND M1A2 AIMING POSTS

2-69. The M14 aiming posts (see figure 2-9) are used to establish an aiming point (reference line) when laying the mortar for deflection. They are made of aluminum tubing and have a pointed tip on one end. Aiming posts have red and white stripes so they can be easily seen through the sight. The M14 aiming post comes in a set of eight segments, plus a weighted stake for every 16 segments to be used as a driver in hard soil.

2-70. The stake has a point on each end, and after emplacement it can be mounted with an aiming post. The segments can be stacked from tip to tail, and they are carried in a specially designed case with a compartment for each segment. Four M1A2 aiming posts are provided with the mortar. They may be stacked up to two high.

Mortar Equipment and Employment

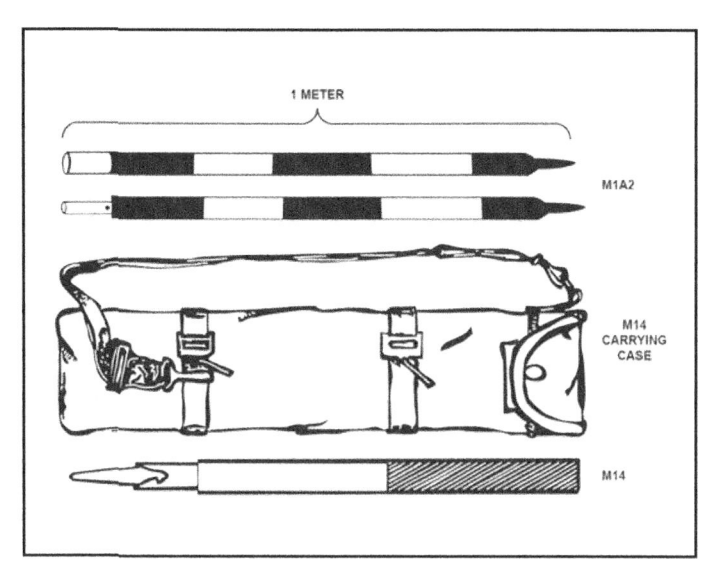

Figure 2-9. M14 and M1A2 aiming posts

M58 AND M59 AIMING POST LIGHTS

2-71. Aiming post lights (see figure 2-10) are attached to the aiming posts so they can be seen at night through the sight. The near post must have a different color light than the far post.

2-72. Aiming post lights come in sets of three—two green (M58) and one orange (M59). A third light is issued for the alternate aiming post. Each light has a clamp, tightened with a wing nut, for attachment to the aiming post. The light does not have a cover for protection when not in use and does not need batteries.

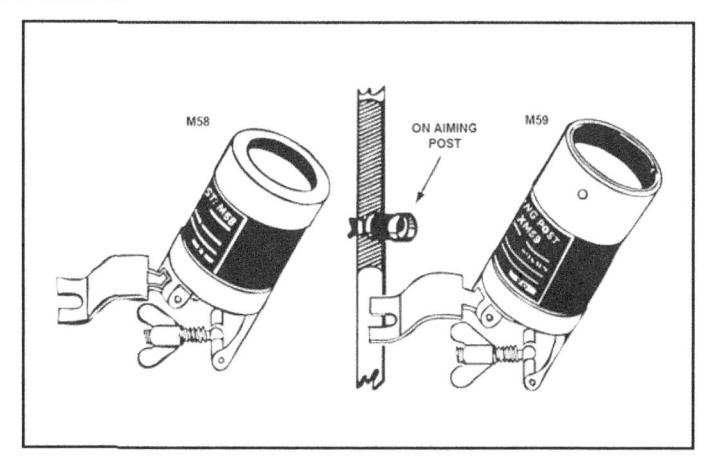

Figure 2-10. M58 and M59 aiming post lights

Chapter 2

> **WARNING**
>
> Radioactive material (tritium gas [H^3]) is used in the M58 and M59 aiming post lights. Radioactive leakage may occur if the M58 and M59 aiming post lights are broken or damaged. If exposed to a broken or damaged M58 or M59 aiming post light or if skin contact is made with any area contaminated with tritium, immediately wash with nonabrasive soap and water, and notify the local chemical, biological, radiological, and nuclear officer, unit chain of command, follow unit SOPs, and local regulations for containment.

M16 AND M19 PLOTTING BOARDS

2-73. M16 and M19 plotting boards are the secondary means of fire control for all forms of digital mortar fire control. The M16 plotting board consists of the base, azimuth disk, and range arm:

- The base is a white plastic sheet bonded to a magnesium alloy backing. The grid system printed on the base is to a scale of 1:12,500, making each small square 50 meters by 50 meters and each large square 500 meters by 500 meters. At the center of the base is the pivot point to which the azimuth disk is attached.
- The clear plastic azimuth disk is roughened on one side so it can be written on with a soft lead pencil. The azimuth scale on the outer edge is numbered every 100 mils (from 0 to 6300) and divided every 10 mils with a longer line at every 50 mils, giving a complete circle of 6400 mils.
- The range arm, a transparent plastic device, has a knob with a pivot pin, two range scales (one on each edge), a protractor on the right bottom, and a vernier scale across the top.

2-74. The M19 plotting board consists of the base, azimuth disk, and range arm:

- The base is a white plastic sheet bonded to a magnesium alloy backing. A grid is printed on the base (in green) at a scale of 1:25,000. The vertical centerline is graduated and numbered up and down (from 0 through 32) from the center (pivot point) in 100s of meters, with a maximum range of 3200 meters. Each small grid square is 100 meters by 100 meters.
- The rotating azimuth disk is made of plastic. Its upper surface is roughened for marking and writing. A mil scale on the outer edge is used for plotting azimuths and angles. It reads clockwise to conform to the azimuth scale of a compass.
- The range arm is used when mortars are plotted at the pivot point. It is made of plastic and can be plugged into the pivot point. Two range scales are on the range arm. On the right edge is a range scale that corresponds to the range scale found on the vertical centerline. An alternative range scale is on the left edge of the range arm and is used when plotting away from the pivot point. The vernier scale at the upper end of the range arm is used to read azimuths or deflection when plotting at the pivot point without rotating the disk back to the vertical centerline.

Note. Refer to FM 3-22.91 for more information on plotting boards.

MFCS OR LHMBC

2-75. The FDC uses the MFCS or LHMBC to convert observer data into fire commands. The MFCS provides a complete, fully integrated, digital, onboard fire control system for the carrier-mounted 120-mm mortar. It provides a shoot-and-scoot capability to the carrier-mounted M121 mortar. The MFCS-Dismounted M-150/M151 operating software is the same as the MFCS and functions identically. For more information on the MFCS-Dismounted System and components of the M120A1 system refer to most updated TMs. With the MFCS, the mortar FDC computes fire commands to execute fire missions and controls its gun tracks. The carrier-mounted MFCS components work together to compute targeting solutions, direct the movement of vehicles into firing positions, allow real-time orientation, and present gun orders to the gunner.

2-76. The LHMBC is a lightweight ruggedized personal digital assistant. It provides the essential functions of the MFCS with similar software in a portable package and allows the operator to quickly calculate accurate ballistic solutions for all current U.S. Army mortar cartridges. The basic model of the LHMBC can be expanded to include Global Positioning System (GPS) and digital communications.

2-77. These commands then are reported to the firing section. The computer uses DA Form 2188-1 (*LHMBC/MFCS Data Sheet*) to record data that pertains to the mortar section or platoon and the firing data for each target engaged when using the LHMBC or MFCS.

Note. Refer to FM 3-22.91 and the appropriate TM for more information on FDC procedures, the MFCS, and the LHMBC.

Chapter 2

LAYING OF THE SECTION

2-78. When all mortars in the section are mounted, the section leader lays the section parallel on the prescribed azimuth with an aiming circle. The mortar section normally fires a parallel sheaf. (See figure 2-11.) To obtain this sheaf, it is necessary to lay the mortars parallel. When a section moves into a firing position, the FDC determines the azimuth on which the section is to be laid and notifies the platoon sergeant (section leader). Before laying the mortars parallel, the section leader must calibrate the mortar sights. All mortars then are laid parallel using the aiming circle, mortar sight, or compass. The section normally is laid parallel by the following two steps:

- Establish the 0-3200 line of the aiming circle parallel to the mounting azimuth.
- Lay the section parallel to the 0-3200 line of the aiming circle (reciprocal laying).

Note. Digital systems, including the MFCS, allows for terrain positioning of the individual mortar guns within the platoon or section. This section discusses the process used by mortars when terrain positioning is not utilized or available.

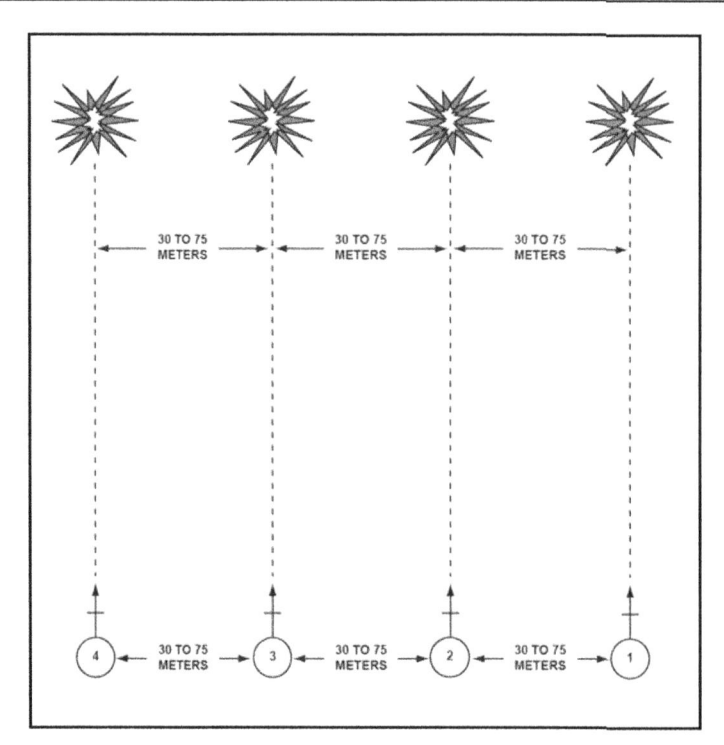

Figure 2-11. Parallel sheaf

Note. The distance between guns is based on caliber of the mortar system.

CALIBRATION FOR DEFLECTIONS WITH AN M2A2 AIMING CIRCLE

2-79. Two methods can be used to calibrate the sight for deflection using the M2A2 aiming circle: the angle method and the distant aiming point (DAP) method. The calibration for deflection using the **angle method** (see figure 2-12) uses the geometric principle that the alternate interior angles formed by a line intersecting two parallel lines are equal. When using this method, the following actions should be taken:

- Set up the aiming circle 25 meters to the rear of the mounted mortar. The mortar is mounted at 800 mils elevation and at a deflection of 3200 mils.
- With the aiming circle fine leveled, index 0 on the azimuth scale and azimuth micrometer scale. Using the orienting motion (nonrecording motion), align the vertical line of the reticle of the telescope so that it bisects the baseplate.
- Traverse and cross-level the mortar until the center axis of the barrel from the baseplate to the muzzle is aligned with the vertical line of the aiming circle telescope reticle.

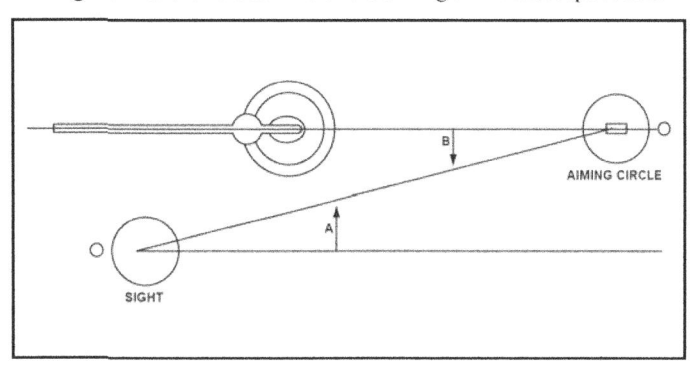

Figure 2-12. Calibration for deflection using the angle method

- Turn the deflection knob of the sight until the vertical line is centered on the lens of the aiming circle and read angle A, opposite the fixed index.
- Turn the azimuth micrometer knob of the aiming circle until the vertical line of the telescope is laid on the center of the sight lens and read angle B, opposite the azimuth scale index. If the sight is in calibration, angles A and B are equal. If they are not equal the sight is adjusted by loosening the two screws in the face of the deflection knob of the sight and slipping the micrometer deflection scale until the scale is indexed at the same reading as angle B of the aiming circle.

Note. The distance between guns is based on the caliber of the mortar system.

2-80. The **DAP method** (see figure 2-13, page 2-22) places the aiming circle and the sight unit on essentially the same line of sight. When using this method, the following actions should be taken:

- Set up the aiming circle and fine level. Align the vertical line of the telescope on a DAP (a sharp, distinct object not less than 200 meters in distance).
- Move the mortar baseplate until the baseplate is bisected by the vertical line of the telescope of the aiming circle. Mount the mortar at an elevation of 800 mils. Traverse and cross-level the mortar until the axis of the barrel from the baseplate to the muzzle is bisected by the

Chapter 2

vertical line of the aiming circle. (The mortar should be mounted about 25 meters from the aiming circle.)

Note. Indexing the aiming circle at 0 is not necessary because only the vertical line is used to align the mortar with the DAP method.

- The aiming circle operator moves to the mortar and lays the vertical line of the sight on the same DAP. If the sight is calibrated, the deflection scales of the sight are slipped to a reading of 3200.

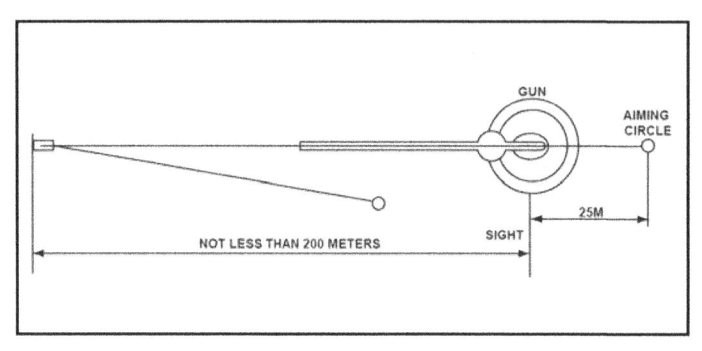

Figure 2-13. Calibration for deflection using the distant aiming point method

RECIPROCAL LAY USING THE MORTAR SIGHTS

2-81. The mortar section can be laid parallel by using the mortar sights (see figure 2-14.). For this method, the mortars should be positioned so that all sights are visible from the base mortar. The base mortar (normally No. 2) is laid in the desired direction of fire by compass or by registration on a known point. After the base mortar is laid for direction, the remaining mortars are laid parallel to the base mortar as follows:

- The section leader moves to the mortar sight of the base mortar and commands, "SECTION, AIMING POINT THIS INSTRUMENT." The gunners of the other mortars refer their sights to the sight of the base mortar and announce, "AIMING POINT IDENTIFIED."

- The section leader reads the deflection from the scale on the sight of the base mortar. He then determines the back azimuth of that deflection and announces it to the other gunners.

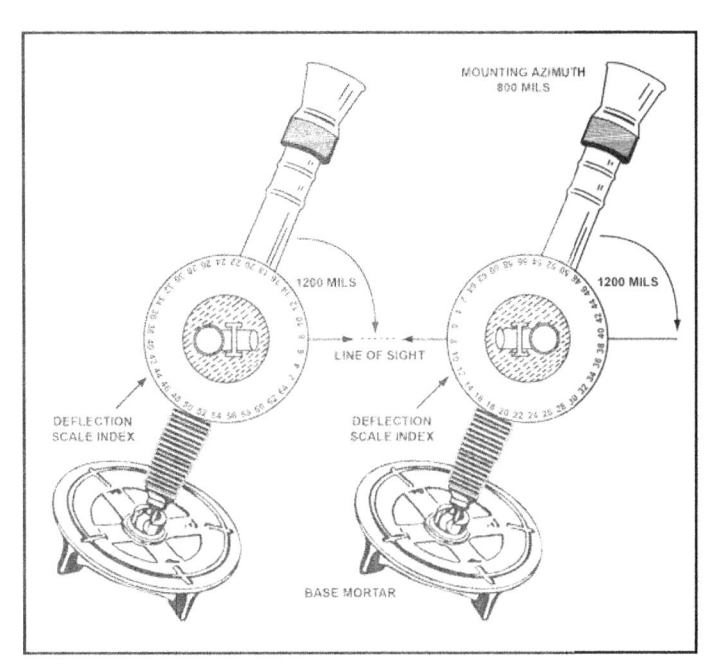

Figure 2-14. Mortar laid parallel with sights

EXAMPLE

Base mortar mounting azimuth: 0800 mils

Base deflection to No. 3 gun: 1200 mils

Add or subtract 3200 mils and announce the reciprocal deflection to the gun to be laid, "NUMBER 3 GUN, DEFLECTION 4400 MILS."

- Each gunner repeats the announced deflection for their mortar, places the deflection on the sight, and lays again on the sight of the base mortar. When the lens of the base mortar sight is not visible, the gunner lays the vertical cross hair of the sight on one of the other three mortar sights (see figure 2-15, page 2-24). He is laid in by this mortar once it is parallel to the base mortar sight unit and then announces, "NUMBER ONE (or THREE), READY FOR RECHECK."

Chapter 2

Figure 2-15. Sighting on the mortar sight

- After each mortar has been laid parallel within 0 (or 1 mil), the mortar barrels are parallel to the base mortar.
- As soon as each mortar is laid, the section leader commands, "NUMBER THREE, REFER DEFLECTION TWO EIGHT ZERO ZERO (2800), PLACE OUT AIMING POSTS."

RECIPROCAL LAY USING THE M2 COMPASS

2-82. A rapid means of laying the section parallel is by using the compass. This is an alternate means and is used only when an aiming circle is not available or when time dictates. It is not as accurate as the previously described methods.

- Before mounting the mortars, each squad leader places a base stake in the ground to mark the approximate location of the mortar.
- The section leader announces the desired mounting azimuth; for example, "MOUNT MORTARS, MAGNETIC AZIMUTH TWO TWO ZERO ZERO."
- Each squad leader places their compass on the base stake marking the location of the mortar and orients the compass on the desired mounting azimuth. By sighting through the compass, he directs the ammunition bearer in aligning the direction stakes along the mounting (magnetic) azimuth.
- Each mortar is mounted and laid on the direction stakes with a deflection of 3200 placed on the sight. The mortar barrels are laid parallel. Once laid for direction, the referred deflection is announced and the aiming posts are emplaced. The direction stakes may be used as the aiming posts but this limits the available amount of mils before the barrel blocks the line of sight.

Note. Because of the differences in individual compasses, the section leader can order that all mortars be laid with one compass. This eliminates some mechanical error. It is possible to lay only the base mortar as described above and then lay the remaining mortars parallel using the mortar sight method.

PLACING OUT AIMING POSTS

2-83. When a firing position is occupied, the section leader must determine which direction the aiming posts are to be placed. Factors to consider are terrain, sight blockage, traffic patterns in the section area, and the mission. After the section leader has laid the section for direction, he commands the section to place out aiming posts on a prescribed referred deflection. The gunner uses the arm-and-hand signals displayed in figure 2-16, page 2-26, to direct the ammunition bearer to place out the aiming posts.

2-84. When the **referred deflection is 2800 mils**, the section leader commands, "SECTION, REFER DEFLECTION TWO EIGHT ZERO ZERO, PLACE OUT AIMING POSTS." If possible, the aiming posts should be placed out to the left front on a referred deflection of 2800 mils. This direction gives large latitude in deflection change before sight blockage occurs. Under normal conditions the front aiming post is placed out 50 meters and the far aiming post 100 meters, for best sight alignment.

2-85. When the **referred deflection is NOT 2800 mils**, the section leader commands, "SECTION, REFER DEFLECTION XXXX, PLACE OUT AIMING POSTS." When local terrain features do not permit placing out the aiming posts at a referred deflection of 2800 mils, the following procedure is used:

- The initial steps are the same as those given above. The gunner refers the sight to the back deflection of the referred deflection and directs the ammunition bearer to place out two aiming posts 50 and 100 meters from the mortar position.
- If the gunner receives a deflection that would be obscured by the barrel, he indexes the referred back deflection, and lays in on the rear aiming post.

2-86. When **any direction fire capability** is desired, two more aiming posts may be placed out to prevent a sight block if the mortar is used in a 6400-mil capability. The section leader then must select an area to the rear of each mortar where aiming posts can be placed on a common deflection for all mortars. A common deflection of 0700 mils is preferred; however, any common deflection to the rear may be selected when obstacles, traffic patterns, or terrain prevent use of 0700 mils.

2-87. If the gunner cannot lay the aiming posts on the prescribed referred deflection, he determines which general direction allows a maximum traverse before encountering a sight block. The gunner notifies the squad leader. The squad leader and FDC determine the actions to be taken.

Note. Units develop SOPs for night signaling of emplacing aiming posts. Based on unit systems, platforms, and equipment, units may direct the use of infrared illumination devices, chemical lights, or filtered flash lights.

Chapter 2

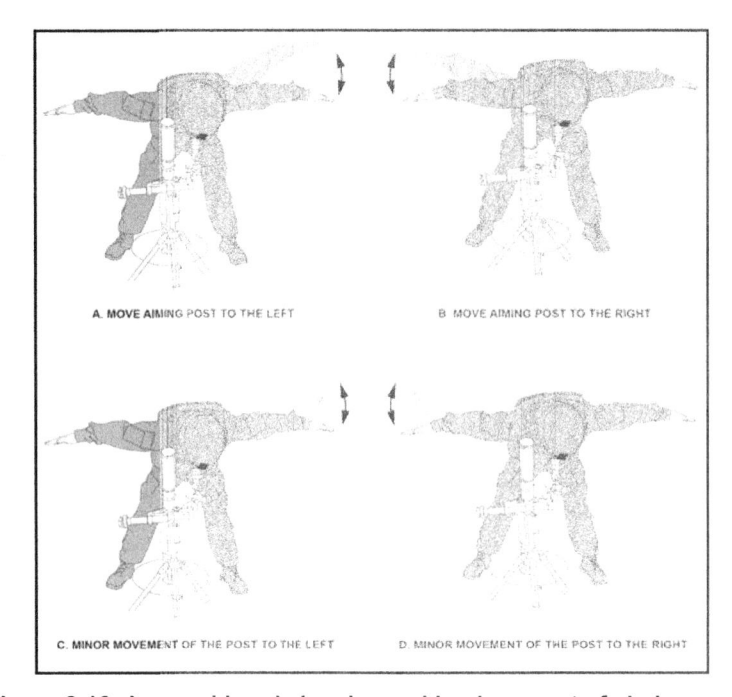

Figure 2-16. Arm-and-hand signals used in placement of aiming posts

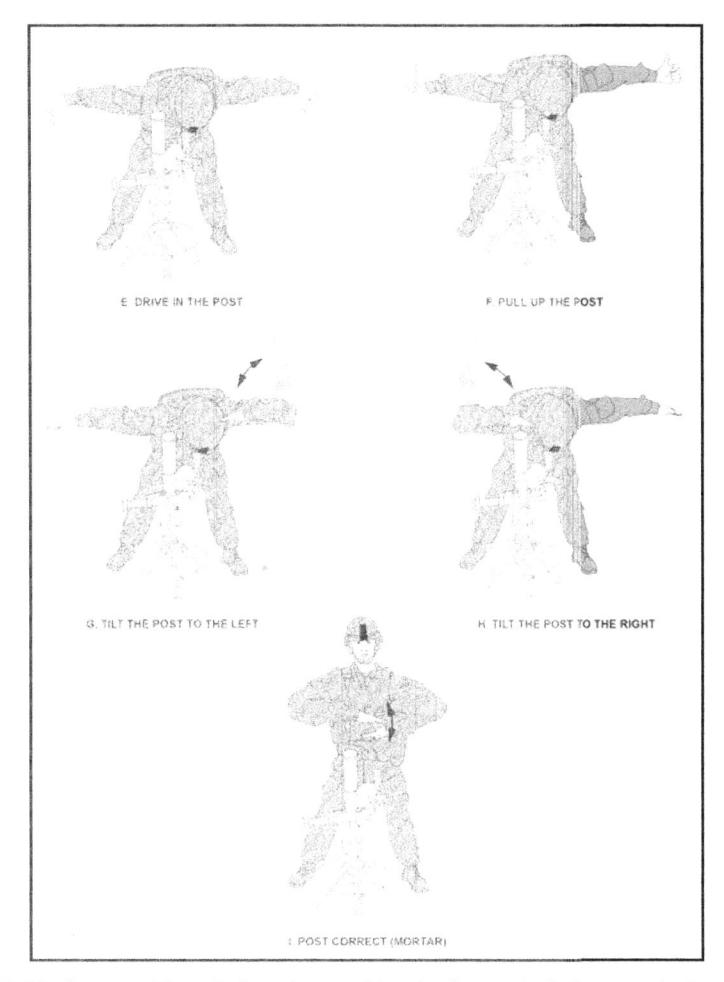

Figure 2-16. Arm-and-hand signals used in placing out aiming posts (continued)

This page intentionally left blank.

Chapter 3

60-mm Mortar

The M224/M224A1 60-mm Lightweight Company Mortar System provides Ranger units, Infantry brigade combat teams and Stryker brigade combat teams with effective and lightweight fire support. Its relatively short range and the small explosive charge can be compensated for with careful planning and expert handling. The mortar can be fired accurately with or without an FDC.

ORGANIZATION AND DUTIES

3-1. This section discusses the organization and duties of the 60-mm mortar squad and section.

Note. Refer to TM 9-1010-233-10 and TM 9-1010-233-23&P for more information on the M224A1 mortar system.

3-2. If the mortar section is to operate quickly and effectively in accomplishing its mission, mortar squad members must be proficient in individually assigned duties. Correctly applying and performing these duties enables the mortar section to perform as an effective fighting team. The section leader commands the section and supervises the training of the elements, using the chain of command to assist in effecting his command and supervising duties.

3-3. The 60-mm mortar squad consists of three Soldiers. Each squad member is cross-trained to perform all duties involved in firing the mortar. The positions and principal duties are as follows.

3-4. The **squad leader** is in position to best control the mortar squad. The squad leader is positioned to the left or rear of the mortar, and facing the cannon. He is also the FDC.

3-5. The **gunner** is on the left side of the mortar in order to manipulate the sight, elevate the gear handle, and traverse the assembly wheel. He places firing data on the sight and lays the mortar for deflection and elevation. Assisted by the squad leader (or ammunition bearer), the gunner makes large deflection shifts by shifting the bipod assembly, and verifying the correct round type, fuze setting, and charge before firing.

3-6. The **ammunition bearer** is to the right rear of the mortar. He prepares the ammunition, assists the gunner in shifting and manipulation of elevation, emplaces aiming posts, loads the mortar, and swabs the cannon every 10 rounds (or after the end of each mission).

COMPONENTS

3-7. The M224/M224A1 60-mm mortar is a muzzle-loaded, smooth-bore, high-angle-of-fire weapon that can be fired in the conventional or handheld mode. It can be drop-fired or trigger-fired and has four major components (see figure 3-1, page 3-2.) These are the components:

Chapter 3

- M225A1 60-MM Mortar Cannon
- M170A1 60-MM Mortar Mount (Bipod)
- M7A1, M8 and M8A1 Baseplates
- M67 or M64A1 Sight Unit

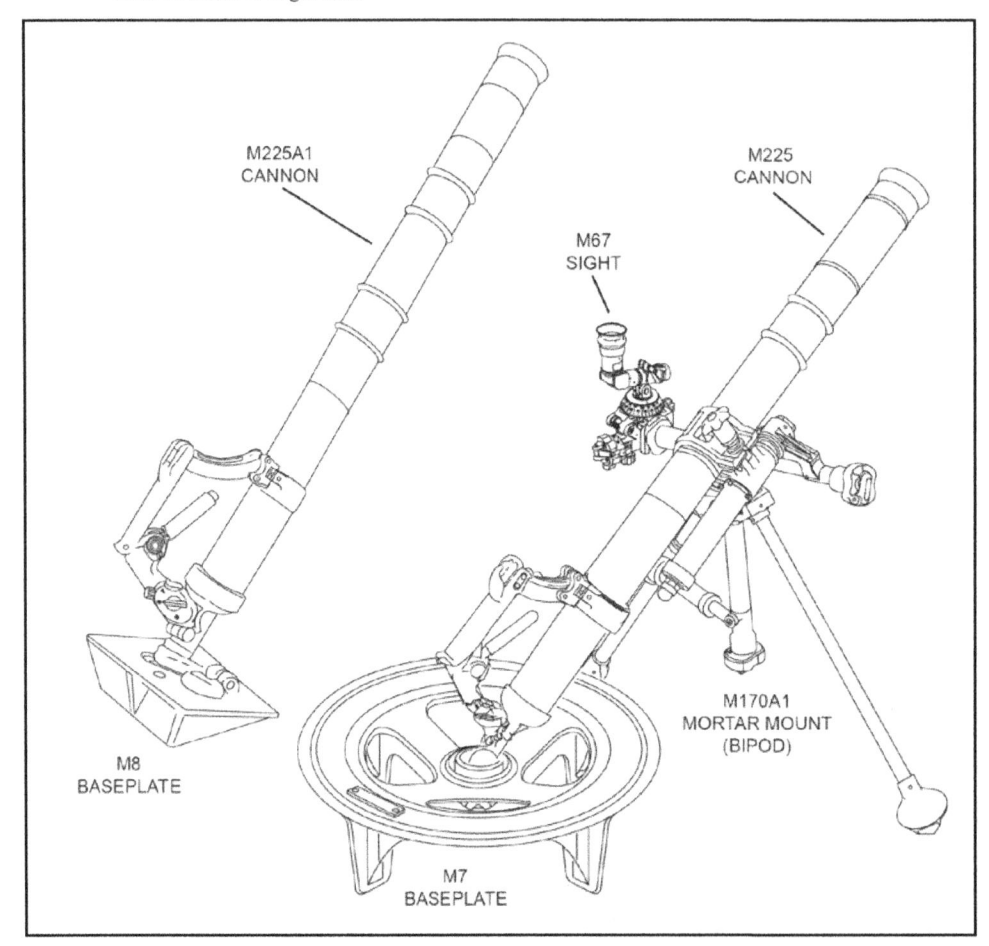

Figure 3-1. M224/M224A1 60-mm mortar

Note. Each mortar cartridge has its own minimum and maximum range. The actual range of any given mortar system is determined by that cartridge which possesses the maximum range.

60-mm Mortar

M225/M225A1 Cannon Assembly

3-8. The cannon assembly (see figure 3-2) has one end closed by a breech cap. The breech cap end of the cannon has cooling fin on the outside that reduce heat generated during firing. Attached to the breech cap end is a combination carrying handle and firing mechanism. The carrying handle has a trigger, firing selector, range indicator, and auxiliary carrying handle. On the outside of the cannon is an upper and a lower firing saddle. The lower saddle is used when firing at elevations of 1100 to 1511 mils; the upper saddle is used when firing at elevations of 0800 to 1100 mils.

Note. When the bipod is positioned in the upper saddle, one turn of the traversing handwheel moves the cannon 10 mils. When the bipod is positioned in the lower saddle, one turn of the traversing handwheel moves the cannon 15 mils.

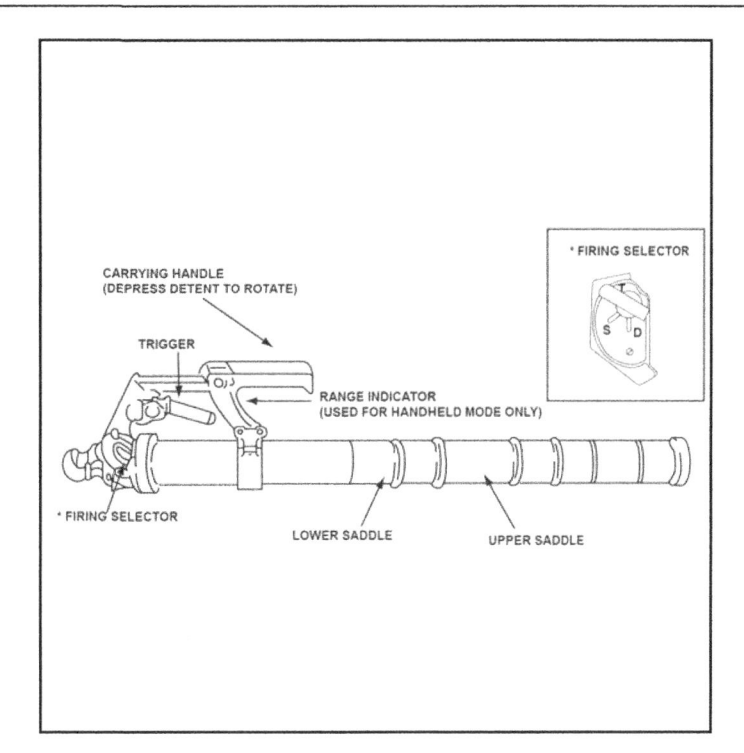

Figure 3-2. M225/M225A1 cannon assembly

M170/M170A1 Bipod Assembly

3-9. The bipod assembly (see figure 3-3, page 3-4) can be attached to the cannon either before or after assembly of the cannon to the baseplate. It consists of seven subassemblies.

3-10. The **collar assembly**, with an upper and lower half, is hinged on the left and secured by a locking knob on the right. The collar fastens in one of the two firing saddles (depending on the elevation being fired), securing the bipod to the cannon. The upper saddle is for the elevations of 1100-1511. The lower saddles is for the elevations of 0800-1100.

Chapter 3

3-11. Two **shock absorbers** located on the underside of the collar assembly protect the bipod and sight from the shock of recoil during firing.

3-12. The **traversing mechanism** moves the collar assembly left or right when the traversing hand crank is pulled out and turned. The hand crank is turned clockwise to move the cannon to the right, and counterclockwise to move the cannon to the left. The left side of the traversing mechanism has a dovetail slot to attach the sight to the bipod.

3-13. The **elevating mechanism** is used to elevate or depress the cannon by turning the hand crank at the base of the elevation guide tube. This assembly consists of an elevating spindle, screw, hand crank, and housing (elevation guide tube). The housing has a latch to secure the collar and shock absorbers to the housing for carrying. The hand crank is turned clockwise to depress, and counterclockwise to elevate.

3-14. The **right leg assembly** consists of a foot, tubular steel leg, a locking latch, and hinge attached to the elevating mechanism housing. The locking latch is located near the hinge and is used to lock the legs in an open or closed position. The **left leg assembly** consists of a foot, tubular steel leg, hinge attached to the elevating mechanism housing, locking nut, and fine cross-leveling sleeve.

3-15. The locking sleeve is near the spiked foot. It is used to lock the elevation housing in place. The **fine cross-leveling nut** above the locking sleeve is used for fine leveling (M170 bipod). The **cross-leveling mechanism** attached to the locking sleeve is used for fine leveling (M170A1 bipod).

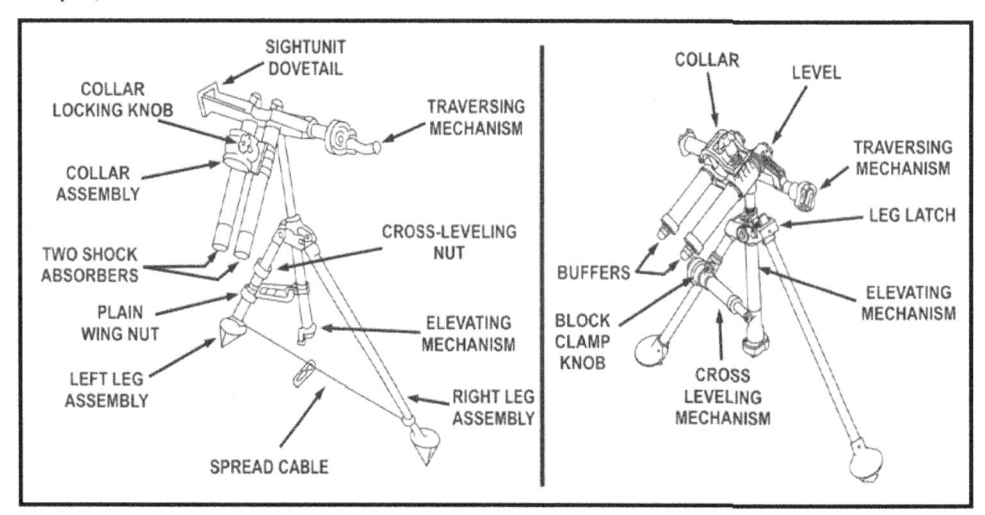

Figure 3-3. M170/M170A1 bipod assembly

M7/M7A1 BASEPLATE

3-16. The M7/M7A1 baseplate (see figure 3-4) is a one-piece, circular, aluminum-forged base. It has a ball socket with a rotating locking cap and a stationary retaining ring held in place by four screws and lock washers. The locking cap rotates 6400 mils, giving the mortar full-circle firing capability. The underside of the baseplate has four spades to stabilize the mortar during firing.

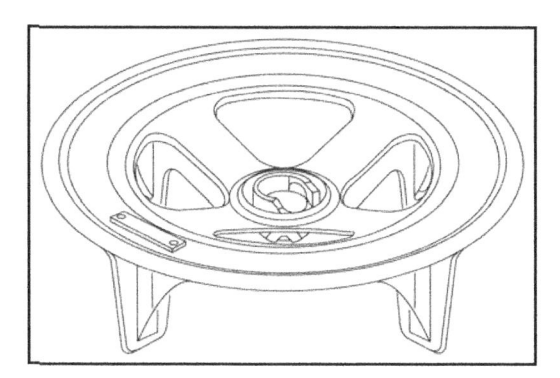

Figure 3-4. M7/M7A1 baseplate

M8/M8A1 BASEPLATE

3-17. The M8/M8A1 baseplate (see figure 3-5) is a one-piece base. It should be used when the mortar is fired in the handheld mode. The baseplate allows the mortar to be fired 0800 mils left and 0800 mils right of the center of sector for a sector coverage of 1600 mils.

3-18. It has a socket where the cannon can be locked to the baseplate by securing the locking arm. The underside of the baseplate has four spades to strengthen and stabilize the mortar during firing. Two spring-loaded plungers lock the baseplate to the cannon in its carry position.

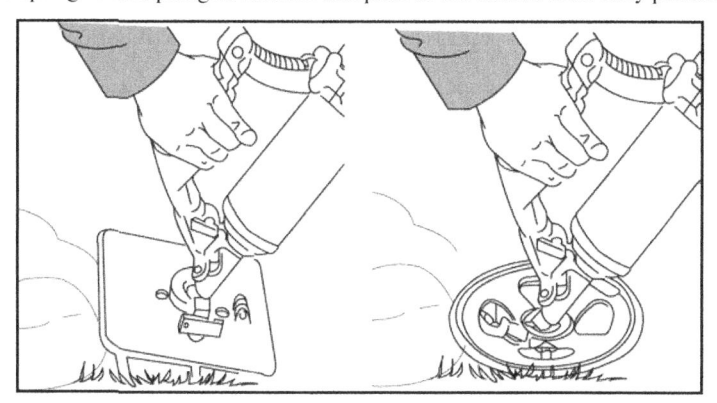

Figure 3-5. M8/M8A1 baseplate

OPERATION

3-19. Safe operation of the 60-mm mortar requires that training include drill practice on tasks for safe manipulation and effective employment. Squad training achieves the speed, precision, and teamwork needed to deliver responsive and effective fire on target.

3-20. Before the mortar is mounted, the squad must perform **premount checks**. Each squad member should be capable of performing all the premount checks.

Squad Leader

3-21. The squad leader performs the premount checks on the cannon so that the—
- Cannon is clean both inside and outside.
- Firing pin is visible.
- Spherical projection is clean, and the firing pin is firmly seated firmly.
- Selector switch is on drop-fire mode.
- Range indicator is not cracked or leaking.

Gunner

3-22. The gunner performs premount checks on the mount so that the—
- Spread cable is fixed to both legs and is taut.
- Clearance on the left leg above the plain wing nut is two fingers in width.
- Locking sleeve is neither too loose nor too tight.
- Traversing bearing is centered.

3-23. When all pieces of equipment are checked, the gunner notifies the section leader by announcing, "ALL CORRECT."

Ammunition Bearer

3-24. The ammunition bearer is responsible for the premount checks on the baseplate ensuring that the—
- Rotatable socket cap moves freely and has a light coat of oil.
- Ribs and braces are checked for breaks and dents, and the inner ring is secured to the outer ring.

MOUNTING OF THE MORTAR

3-25. The squad mounts the mortar by taking the following actions:
- The squad leader picks up and places the sight case and two aiming posts at the exact position where the mortar is to be mounted.
- The ammunition bearer places the outer edge of the baseplate against the baseplate stake, aligning the left edge of the cutout portion of the baseplate with the right edge of the baseplate stake. He then rotates the socket cap so that the open end points in the direction of fire.
- The gunner picks up the bipod using the left hand on the traversing hand crank and right hand on the dovetail slot. He moves forward of the baseplate about 12 to 15 inches and faces the baseplate on line with the left edge (gunner's viewpoint) of the baseplate. Dropping down on one knee in front of the bipod, the gunner supports the bipod with his left hand on the gear case, then detaches the hook and unwraps the cable assembly. The gunner places his left hand on the midsection of the traversing slide and right hand on the mechanical leg, and extends the bipod legs the length of the cable assembly. The gunner then aligns the center of the bipod assembly with the center of the baseplate, and ensures that the elevation guide cannon is vertical and the locking nut is hand-tight. The gunner moves to the mechanical leg side and supports the bipod with the left hand on the shock absorber, and unscrews the collar locking knob to open the collar.

60-mm Mortar

- The ammunition bearer picks up the cannon and inserts the spherical projection of the base plug into the socket, then rotates the cannon 90 degrees to lock it to the baseplate. If performed properly, the carrying handle is on the upper side of the cannon, facing skyward.
- The gunner pushes down on the shock absorber and raises the collar assembly. The ammunition bearer lowers the cannon and places the lower saddle on the lower part of the collar. The gunner closes the upper part of the collar over the cannon, and replaces the locking knob to its original position and makes it hand-tight. The ammunition bearer winds the elevation hand crank up 15 to 17 turns.
- The gunner takes the sight out of the case and sets a deflection of 3200 mils and an elevation of 1100 mils, then mounts the sight to the mortar by pushing the lock latch on the sight inward. The gunner slides the dovetail on the sight into the dovetail slot on the bipod until firmly seated and releases the latch. He should tap up on the bottom of the sight to ensure proper seating. Then, the gunner levels the mortar first for elevation 1100 mils, and then cross-levels. The gunner announces, "(gun number) UP," to the squad leader.

Safety Checks Before Firing

3-26. The entire squad performs safety checks. The gunner ensures that—

- The cannon is locked to the baseplate and the open end of the socket cap points in the direction of fire. The bipod should be connected to either the upper or lower saddle of the cannon.
- The cannon is locked on the collar by the locking knob.
- The locking nut is hand-tight.
- The cable is taut (M170 bipod only).
- The selector switch on the cannon is on drop-fire.
- Check for mask and overhead clearance.

WARNING

Keep feet clear of the mortar baseplate, as a **rapidly** seating baseplate can cause serious injury.

Note. Until the baseplate is seated firmly, remove the sight from the mortar before firing each round.

Mask and Overhead Clearance

3-27. Since the mortar normally is mounted in defilade, there could be **mask** (intervening object which screens the guns from the view of the enemy or target) such as a hill, trees, buildings, or a rise in the ground. Roofs or overhanging tree branches can cause **overhead** interference. The gunner must be sure the round does not strike any obstruction.

3-28. When selecting the exact mortar position, the squad leader quickly checks for mask and overhead clearance. After the mortar is mounted, the gunner checks it thoroughly and determines mask and overhead clearance by sighting along the top of the cannon with his eye placed near the base plug. If the line of sight clears the mask, it is safe to fire. If not, he still may fire at the desired

17 March 2017   TC 3-22.90   3-7

Chapter 3

range by selecting a charge zone having a higher elevation for that particular range. When firing under the control of an FDC, the gunner reports to the FDC that mask clearance cannot be obtained at a certain elevation.

3-29. Firing is slowed if mask clearance is checked before each firing. Therefore, if the mask is not regular throughout the sector of fire, the minimum mask clearance is determined to eliminate the need for checking on each mission. To do this, the gunner depresses the cannon until the top of the mask is sighted. He then levels the elevation bubble and reads the setting on the elevation scale and elevation micrometer. That setting is the minimum mask clearance. The gunner notifies the squad leader of the minimum mask clearance elevation. Any target that requires that elevation or lower cannot be engaged from that position.

3-30. Placing the mortar in position at night does not relieve the gunner of the responsibility for checking for mask and overhead clearance.

3-31. The gunner ensures that the bore is clean and swabs the bore dry. The ammunition bearer ensures that each round is clean, safety pin is present (if applicable), and ignition cartridge is in proper condition to include serviceability and accountability of all propelling charges.

SMALL DEFLECTION AND ELEVATION CHANGES

3-32. With the mortar mounted and the sight installed, the gunner makes small deflection and elevation changes by taking the following actions:

- The gunner lays the sight on the two aiming posts (placed out 50 and 100 meters from the mortar or METT-TC dependent) on a referred deflection of 2800 mils and an elevation of 1100 mils. The mortar is within two turns of center of traverse. The vertical cross hair of the sight is on the left edge of the aiming post with both bubbles leveled within the outer red lines.
- The gunner is given a deflection change in a fire command between 20 and 60 mils. The elevation change announced must be less than 90 mils and more than 35 mils.
- As soon as the sight data are announced, the gunner places it on the sight, lays the mortar for elevation, and then traverses onto the aiming post by turning the traversing handwheel and the adjusting nut in the same direction. A one-quarter turn on the cross-leveling nut equals one turn of the traversing handwheel. When the gunner is satisfied with the sight picture he announces, "UP."

Note. The squad repeats all elements given in the fire command.

- After the gunner has announced "UP," the mortar should be checked by the squad leader to determine if the exercise was performed correctly.

LARGE DEFLECTION AND ELEVATION CHANGES

3-33. With the mortar mounted and the sight installed, the squad makes large deflection and elevation changes by taking the following actions:

- The gunner lays the sight on the two aiming posts (placed out 50 and 100 meters from the mortar or METT-TC dependent) on a referred deflection of 2800 mils and an elevation of 1100 mils.
- The gunner is given a deflection and elevation change in a fire command causing the gunner to shift the mortar between 200 and 300 mils and an elevation change between 100 and 200 mils.

60-mm Mortar

- As soon as the sight data are announced, the gunner places it on the sight, elevates the mortar until the elevation bubble floats freely, and then centers the traversing bearing. If the elevation is between 1100 to 1511 mils, the cannon is mounted in the lower saddle. If the elevation is between 0800 to 1100 mils, the high saddle is used. If the saddle is changed, the squad leader helps the gunner.

- The ammunition bearer moves into position to the front of the bipod on his right knee and grasps the bipod legs (palms up), lifting until they clear the ground enough to permit lateral movement. The gunner moves the mortar as the ammunition bearer steadies it. The ammunition bearer tries to maintain the traversing mechanism on a horizontal plane. The gunner moves the mortar by placing his right hand over the clamp handle assembly and his left hand on the bipod leg. He pushes or pulls the bipod in the direction desired until the vertical cross hair is within 20 mils of the aiming posts. When the approximate alignment is completed, the gunner signals the ammunition bearer to lower the bipod by pushing down on the mortar. (See figure 3-6.)

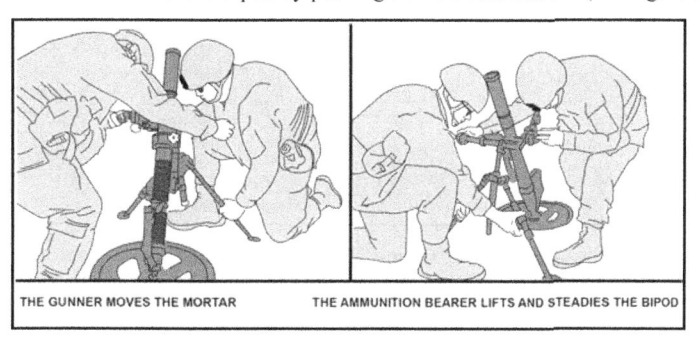

Figure 3-6. Large deflection and elevation changes

- The gunner levels the mortar for elevation, then cross-levels. He continues to traverse and cross-level until the correct sight picture is obtained with both bubbles leveled within the outer red lines. The mortar should be within two turns of center of traverse when the exercise is completed.

REFERRING OF THE SIGHT/REALIGNMENT OF AIMING POSTS

3-34. Referring the sight and realigning aiming posts ensure that all mortars are set on the same data. The section leader, acting as the FDC, has one deflection instead of two. Take the following actions:

- The sheaf is paralleled, and each mortar is laid on the correct data.

- The section leader prepares an administrative announcement using the format for a fire command and the hit data of the base piece as follows:
  - "SECTION."
  - "DO NOT FIRE."
  - "REFER DEFLECTION TWO EIGHT ZERO ZERO (2800)."
  - "REALIGN AIMING POSTS."

- The gunners refer their sights to the announced deflection. Each gunner checks their sight picture. If the sight picture is aligned, no further action is required.

- In laying the mortar for direction, the two aiming posts do not always appear as one when viewed through the sight. This separation is caused by a large deflection shift of the cannon or a rearward displacement of the baseplate assembly (caused by the shock of firing). When the

Chapter 3

aiming posts appear separated, the gunner cannot correctly use either one of them as the aiming point. To lay the mortar correctly, he establishes a compensated sight picture (see figure 3-7) and traverses the mortar until the sight picture appears with the left edge of the far aiming post, which is placed exactly midway between the left edge of the near aiming post and the vertical line of the sight. This corrects for the displacement.

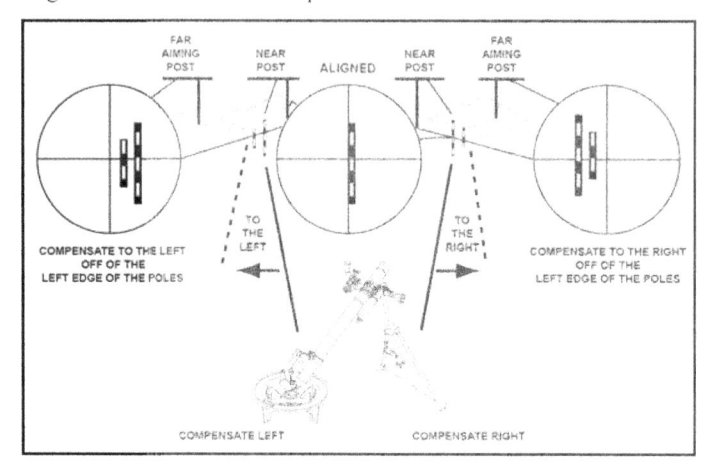

Figure 3-7. Compensated sight picture

- At the first lull in firing, the gunner must determine whether the displacement is caused by traversing the mortar or by displacement of the baseplate assembly. To do this, he places the referred initial deflection on the sight and lays on the aiming posts. If both aiming posts appear as one, the separation is caused by traversing. In this case, the gunner continues to lay the mortar as described and does not realign the aiming posts. When the posts still appear separated, the separation is caused by displacement of the baseplate assembly. The gunner notifies the squad leader, who in turn requests permission from the section leader to realign the aiming posts. To realign the aiming posts using the sight unit, the gunner—
  - Places on the sight the deflection originally used to place out the posts.
  - Lays the mortar so that the vertical line of the sight is aligned on the left edge of the far aiming post.
  - Without shifting the mortar, refers the sight until the vertical cross hair falls on the left edge of the near aiming post. This actually measures the angle between the posts.
  - With this last deflection set on the sight, lays the mortar again until the vertical cross hair is aligned on the far aiming post.
  - Without shifting the mortar, refers the sight again to the original referred deflection used to place out the aiming posts. The line of sight, through the sight, now is parallel to the original line established by the aiming posts.
  - Looking through the sight, directs the ammunition bearer to move the aiming posts so that they are realigned with the sight's vertical line. Now the posts are realigned to correct the displacement.

Note. This procedure is used only when the amount of displacement makes it difficult to obtain a compensated sight picture.

MALFUNCTIONS

3-35. Mortarmen must be aware of misfires, hangfires, cook offs, and short rounds. A **misfire** is a complete failure to fire. It can be caused by a faulty firing mechanism or faulty element in the propelling charge explosive train. A misfire cannot be immediately distinguished from a delay in functioning of the firing mechanism or from a hangfire; therefore, it must be handled with care.

3-36. All firing malfunctions should be considered a misfire. Mechanical malfunctions can be caused by a faulty firing pin or by rounds lodged in the cannon because of burrs, excess paint, oversized rounds, or foreign matter in the cannon.

Note. Procedures for removing a misfire are discussed in the appropriate TM for the mortar system and platform used.

3-37. A **hangfire** is caused by a round becoming lodged in the breach preventing the rounds primer from striking the firing pin at the time of firing. In most cases, the delay ranges from a split second to several minutes. Thus, a hangfire cannot be distinguished from a misfire.

3-38. A **cook off** is a functioning of one or more of the explosive components of a round chambered in a hot weapon, initiated by the heat of the weapon.

3-39. A **short round** occurs when there is incomplete firing of the propelling charge explosive train, causing the round to be discharged under significantly reduced pressure. To avoid erratic or short rounds ensure that all components are free of sand, mud, moisture, frost, snow, ice, or other foreign matter before loading cartridge into cannon.

MODES OF FIRING THE 60-MM MORTAR

3-40. The 60-mm mortar can be fired in two different ways: conventional (with sight, bipod and baseplate) or handheld. Conventional is the preferred method of operation to ensure accuracy.

3-41. The proper procedures for loading and firing a 60-mm mortar in **conventional mode** are the—

- FDC issues a fire command to the squad leader.
- Squad leader records and issues the fire command to the squad.
- Squad repeats the fire command.
- Ammunition bearer prepares the round according to the fire command.
- Squad leader inspects the round before it is passed to the ammunition bearer. The ammunition bearer holds the round with both hands (palms up) near each end of the round body (not on the fuze or the charges).

Note. Communication among the mortar squad is imperative in order to mitigate the probability of human error when selecting the appropriate round type, number of charges, and fuze setting. Whenever possible, the squad leader physically inspects each round prior to firing. If the situation does not allow for physical inspection, the squad leader must visually inspect the round and verbally confirm and acknowledge with the ammunition bearer that the round or rounds are all correct. The squad leader or gunner remains the last control measure to verify that all data for the fire mission is without error (deflection and elevation, sight picture, and round).

Chapter 3

- The ammunition bearer checks the round for correct charges, fuze tightness, and fuze setting.
- When both the gun and the round(s) have been determined safe and ready to fire, the squad leader gives the following command to the FDC: "NUMBER (number of mortar) GUN, UP."
- The ammunition bearer holds the round with the fuze pointed to his left. By pivoting his body to the left, the ammunition bearer accepts the round from the ammunition bearer with his right hand under the round and the left hand on top of the round.
- Once the ammunition bearer has the round, he keeps two hands on it until it is fired.

> **CAUTION**
> The ammunition bearer is the only member of the mortar squad who loads and fires the round.

- The squad leader commands, "HANG IT, FIRE" according to the method of fire given by the FDC.

> **WARNING**
> **Keep feet clear of the mortar baseplate, as a rapidly-seating baseplate can cause serious injury.**

Note. Remove the sight from the mortar before firing until the baseplate is seated firmly.

- The ammunition bearer holds the round in front of the muzzle at about the same angle as the cannon. The gunner conducts final verification that the appropriate charge is set on rounds that have been previously handled and repackaged. The potential for charges to fall off exists, especially when being handled in the limited confines of a vehicle. At the command, "HANG IT," the ammunition bearer guides the round into the cannon tail end first, to a point beyond the narrow portion of the body (about three-quarters of the round) being careful not to hit the primer or charges or disturb the lay of the mortar.
- Once the round is inserted into the cannon the proper distance, the ammunition bearer shouts, "NUMBER (number of mortar) GUN, HANGING."
- At the command, "FIRE," the ammunition bearer releases the round by pulling both hands down and away from the outside of the cannon. The ammo bearer ensures that his hands do not cross the muzzle of the cannon as he drops the round.
- Once the round is released, the gunner and ammunition bearer take a full step toward the rear of the weapon while pivoting their bodies so that they are both facing away from the blast.
- The ammunition bearer pivots to his left and down toward the ammunition bearer, ready to accept the next round to be fired (unless major movements of the bipod require him to lift and clear the bipod off the ground).
- Subsequent rounds are fired based on the FDC fire commands.
- The ammunition bearer ensures the round has fired safely before attempting to load the next round.

- The ammunition bearer does not shove or push the round down the cannon. The round slides down the cannon under its own weight, strikes the firing pin, ignites, and fires.
- The ammunition bearer and gunner, as well as the remainder of the mortar squad, keep their upper body below the muzzle until the round fires to avoid muzzle blast.
- During an FFE, the gunner tries to level all bubbles between each round, ensuring his upper body is away from the mortar and below the muzzle when the ammunition bearer announces, "HANGING," for each round fired.
- The ammunition bearer informs the squad leader when all rounds for the fire mission are expended, and the squad leader informs the FDC when all of the rounds are completed. For example, "NUMBER TWO GUN, ALL ROUNDS COMPLETE."

HANDHELD MODE

3-42. The duty requirements when firing the 60-mm mortar in **handheld mode** remain unchanged from conventional mode, with minor adjustments to techniques. Unit SOPs, policies, and requirements should dictate minimum safe levels of leadership participation when validating these SOPs.

3-43. In the handheld mode, the squad leader calls the range to the target. The squad leader is the FDC and computer and has the final decision on adjustments. The squad leader makes adjustment corrections and observes effects of fire, in some tactical situations. The squad leader performs safety checks, mask, and overhead clearance. The squad leader commands the mortar out of action when mission is complete.

DANGER
TO AVOID PERSONNEL INJURY, FIRE ONLY AT CHARGE 0 OR 1 IN HANDHELD MODE, WITH EITHER BASEPLATE. SEEK MEDICAL ATTENTION IF INJURY OCCURS.

3-44. The gunner seats the baseplate (with force) in the firing position and uses his thumb to align the mortar onto the target, ensuring the thumb remains clear of the barrel. The gunner then checks the range indicator assembly (see figure 3-8, page 3-14) to index the range. The range indicator assembly contains a red scale (item 1, figure 3-8) is used for charge zero, a black scale (item 2, figure 3-8) is used for charge one. The range indicator assembly also contains a ball (item 3, figure 3-8) that moves when the mortar is elevated or depressed and indicates the firing range. The gunner sets the mortar from safe to trigger fire mode, then based on the situation and mission requirements, may set the mortar to drop fire mode. The gunner makes elevation and deviation corrections by movement of the range indicator assembly and alignment of the barrel. The gunner may be responsible for loading the mortar and making adjustments in some tactical situations. The gunner performs safety checks, mask, and overhead clearance. The gunner observes effects and battle damage assessment, when possible. The gunner takes the mortar out of action when the mission complete.

3-45. The M224A1 has a selector knob (see figure 3-9, page 3-15) at the bottom of the range indicator which rotates between HE0, HE1, EF0, and EF1. HE0 is the range for M720 and M888 cartridges at charge 0. HE1 is the range for M720 and M888 cartridges at charge 1. EF0 is the range for M1061 cartridge at charge 0 and EF1 is the range for M1061 cartridge at charge 1.

Chapter 3

> **WARNING**
>
> Only one Soldier may load the cannon at a time. Loading a mortar weapon with two alternating Soldiers can be very dangerous and could prove fatal. Even with one man loading, double loading can occur. This is especially true in rapid-fire exercises. For this reason, it is imperative that there is absolute certainty that the previous round has left the mortar tube before a new round is dropped in. The commander (or the appointee) should keep careful watch for misfire and double loading.

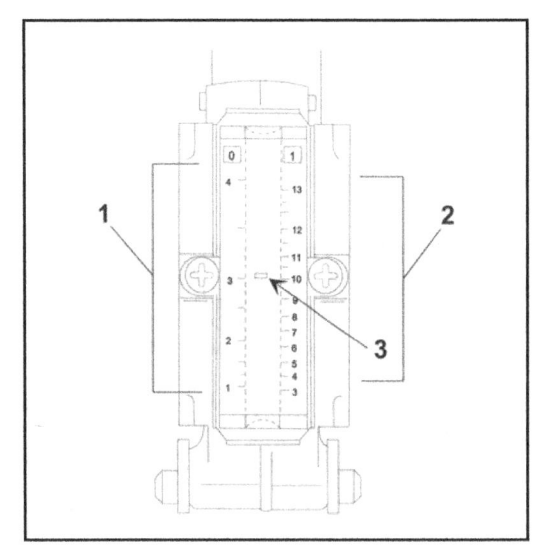

Figure 3-8. M224/M224A1 range indicator assembly

60-mm Mortar

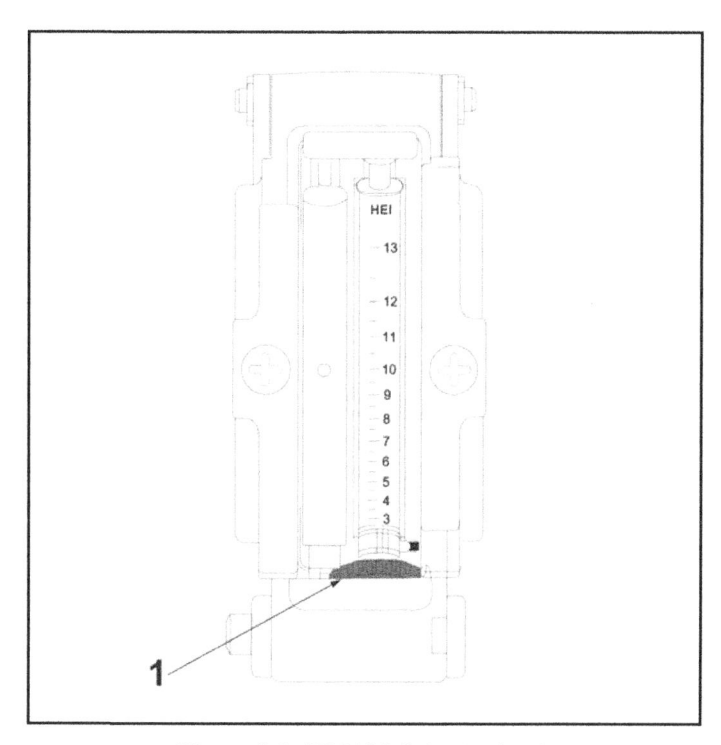

Figure 3-9. M224A1 Selector knob

3-46. The ammunition bearer feeds the rounds while the gunner keeps the gun aligned and makes adjustments to the mortar fire mission. The ammunition bearers' duty remains unchanged from the conventional mode.

> **WARNING**
>
> Keep knees clear of the base plate. Keep thumb clear of the barrel.

DISMOUNTING AND CARRYING THE MORTAR

3-47. To dismount and carry the mortar, the squad leader commands, "OUT OF ACTION." Squads dismount using the following procedures:

- The ammunition bearer retrieves the aiming posts. The gunner removes the sight and places it in the case (use deflection and elevation from TM for storage). Then the ammunition bearer lowers the mortar to its minimum elevation and backs off one-quarter turn, centers the traversing mechanism, and unlocks the collar with the collar locking knob.
- The gunner grasps the base of the cannon and turns it 90 degrees (a one-quarter turn), until the spherical projection is in the unlocked position in the baseplate socket and then lifts up on

the base end of the cannon, removing it from the collar assembly. The ammunition bearer secures the baseplate.

- The gunner (or ammunition bearer) relocks the collar with the collar locking knob, moves to the front of the bipod and faces it. Kneeling on his right knee with the left hand on the gear case, and loosening the locking nut. He tilts the bipod to the left and closes the bipod legs, placing the cable around the legs and rehooking the cable. He stands up, placing the right hand on the sight slot and left hand on the traversing handwheel.
- On command, squad members take the equipment distributed to them by the squad leader and move.

3-48. The mortar can be carried by one or two Soldiers for short distances. Adhere to the following procedures to properly carry the mortar:

- For a one-man carry, the mortar is in the firing position with the mount attached to the cannon. The elevating mechanism is fully depressed, and the bipod legs are together. The mount is folded back underneath the cannon. The carrying handle is used to carry the complete mortar.
- For a two-man carry, the M7 baseplate with aiming post bag and sectional artillery staff is one load, and the cannon/mount combination with M8 baseplate and sight unit is the second load. The mount is attached to the cannon, and the elevating mechanism is depressed fully. The bipod legs are together, and the bipod is folded up under the cannon.
- For the handheld mode, the M8 baseplate is left attached to the cannon. The baseplate is rotated 90 degrees to the right and rotated up until the two spring plungers on the front edge of the baseplate body latch onto the protrusion on the right side of the basecap. Then the auxiliary carrying handle is placed in the carrying position.

The 60-mm section leader and squad leader carry all necessary equipment to function as an FDC.

Chapter 4

81-mm Mortar

The M252/M252A1 81-mm mortar delivers timely, accurate fires to meet the requirements of supported companies and troops in the Ranger, Infantry Brigade Combat Team (IBCT), ABCT, and Stryker Brigade Combat Team (SBCT) formations. The 81-mm mortar is lightweight enough to maneuver through challenging terrain while providing accurate indirect fire with or without and FDC. This chapter discusses assigned personnel duties, squad drill, mechanical training, and characteristics of the mortar.

ORGANIZATION AND DUTIES

4-1. Each member of the Infantry mortar squad has principle duties and responsibilities. (Refer to ATTP 3-21.90 for a discussion of the duties of the platoon headquarters.)

Note. Refer to TM 9-1015-257-10 or TM 9-1015-257-23&P for M252A1 81-mm mortar. Refer to TM 9-1015-249-10 for the legacy M252 mortar.

4-2. If the mortar squad and section are to operate quickly and effectively in accomplishing their mission, mortar squad members must be proficient in individually assigned duties. Correctly applying and performing these duties enables the mortar section to perform as an effective fighting team. The platoon leader commands the platoon and supervises the training of the elements. He uses the chain of command to assist in effecting his command and supervising duties.

4-3. The mortar squad consists of four Soldiers (see figure 4-1, page 4-2)—
- Squad leader.
- Gunner.
- Assistant gunner.
- Ammunition bearer.

4-4. The **squad leader** is positioned behind the mortar to command and control the squad. He supervises the emplacement, laying, and firing of the mortar, and all other squad activities. The squad leader also verifies the correct round type, fuze setting, and charge prior to firing.

4-5. The **gunner** is positioned to the left side of the mortar to manipulate the sight, elevating handwheel, and traversing handwheel. He places firing data on the sight and lays the mortar for deflection and elevation. The gunner makes large deflection shifts by directing the assistant gunner to shift the bipod assembly while the gunner keeps the bubbles level during firing.

4-6. The **assistant gunner** stands to the right of the mortar, facing the cannon and ready to load. In addition to loading, he swabs the bore after every 10 rounds or after each fire mission. The

Chapter 4

assistant gunner may assist the gunner in shifting the mortar when the gunner is making large deflection changes.

4-7. The **ammunition bearer** stands to the right rear of the mortar. He maintains the ammunition for firing, prepares the ammunition and passes it to the assistant gunner, and provides local security for the mortar position. The ammunition bearer may also act as the squad driver.

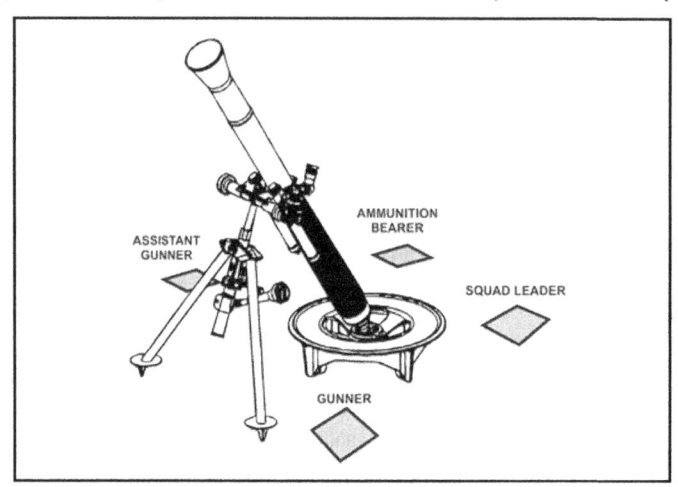

Figure 4-1. Position of squad members

COMPONENTS

4-8. The M252/M252A1 81-mm mortar is a smooth-bore, muzzle-loaded, high angle-of-fire weapon. The components of the mortar consist of a cannon, mount, and baseplate. This section discusses the characteristics and nomenclature of each component (see figure 4-2).

Note. The M252A1 81-mm mortar is scheduled to replace the M252.

81-mm Mortar

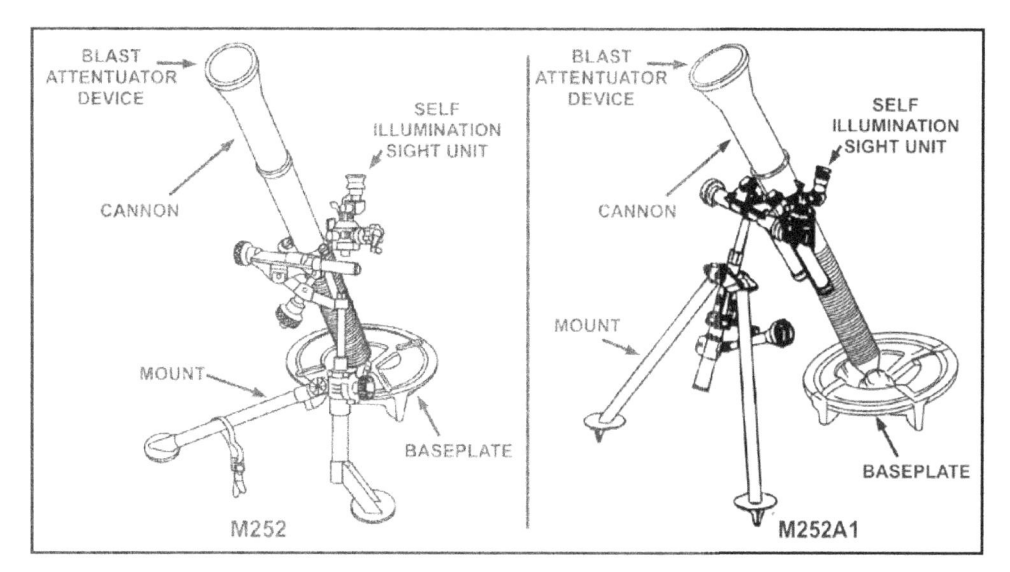

Figure 4-2. M252/M252A1 81-mm mortar

Note. Each mortar cartridge has its own minimum and maximum range. The actual range of any given mortar system is determined by that cartridge which possesses the maximum range.

4-9. The **M253 cannon assembly** consists of the cannon that is sealed at the lower end with a removable breech plug, which houses a removable firing pin (see figure 4-3). At the muzzle end is a cone-shaped blast attenuation device (BAD) that is fitted to reduce noise. The BAD is removed only by qualified maintenance personnel.

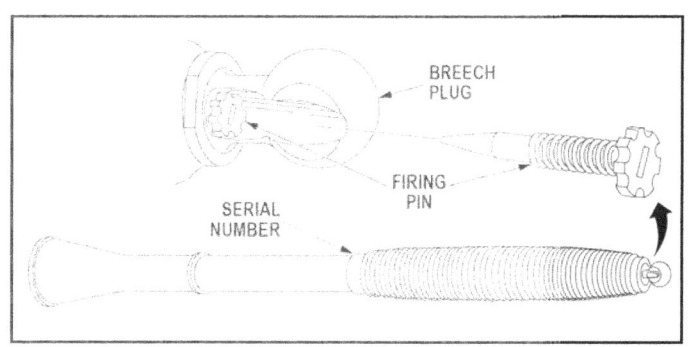

Figure 4-3. M253 cannon assembly

Chapter 4

4-10. The **M177 mount** consists of elevating and traversing mechanisms and a bipod. (See figure 4-4.)

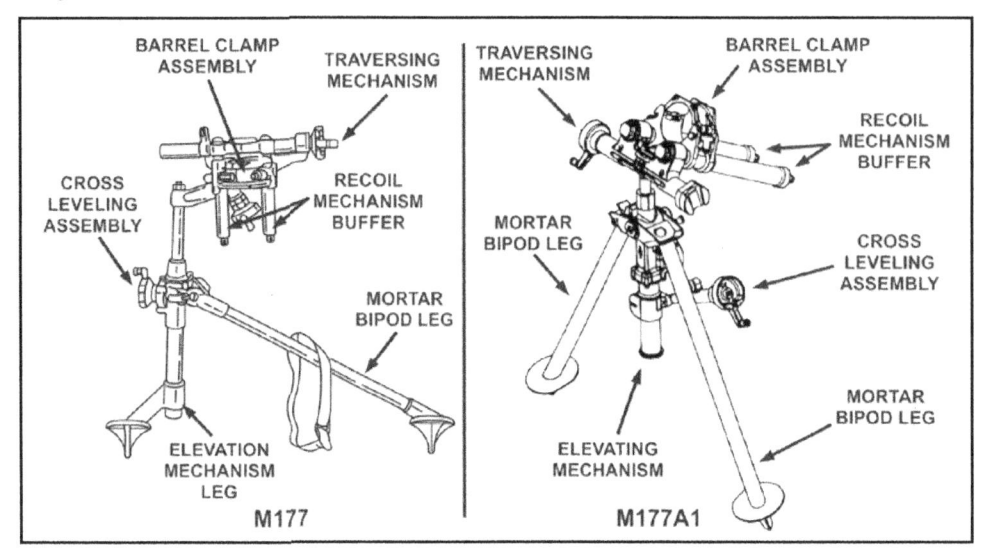

Figure 4-4. M177/M177A1 mount in folded position

4-11. The **bipod** provides front support for the cannon and carries the gears necessary to lay the mortar. The cannon clamp, consisting of an upper and lower clamp, is situated at the top. The upper clamp is fitted with a locking arrangement that is made up of a curved handle and a spring-loaded locking rod that is ball-shaped at its lower end. The lower clamp is shaped and bored on each side to house the buffer cylinders. On the right side, the clamp is recessed to receive the ball end of the locking rod. A safety latch located at the side of the recess is used to secure the ball.

4-12. The **baseplate** (see figure 4-5) is constructed of one piece and supports and aligns the mortar for firing. During firing, the breech plug on the cannon is sealed and locked to the rotatable socket in the baseplate.

TRAVERSING MECHANISM

4-13. The sight bracket is attached to the buffer carrier, which is fitted to the traversing screw assembly. Attached to the right of the screw is the traversing handwheel. The traversing screw assembly is fitted to the clamp assembly, which is pivoted in the center on an arm attached to the elevating leg. Attached to the arm is the cross-leveling mechanism, which is attached to the clamp assembly at its upper end.

ELEVATING MECHANISM

4-14. The elevating shaft is contained in the elevating leg; to the left of the elevating leg is the elevating handwheel. A plain leg is fitted to a stud on the elevating leg and is secured by a leg-locking handwheel. A spring-loaded locating catch is behind the elevating gear housing, which locates the plain leg in its supporting position for level ground. A securing strap is attached to the plain leg for securing the bipod in the folded position. Both legs are fitted with a disk-shaped foot with a spike beneath to prevent the mount from slipping.

81-mm Mortar

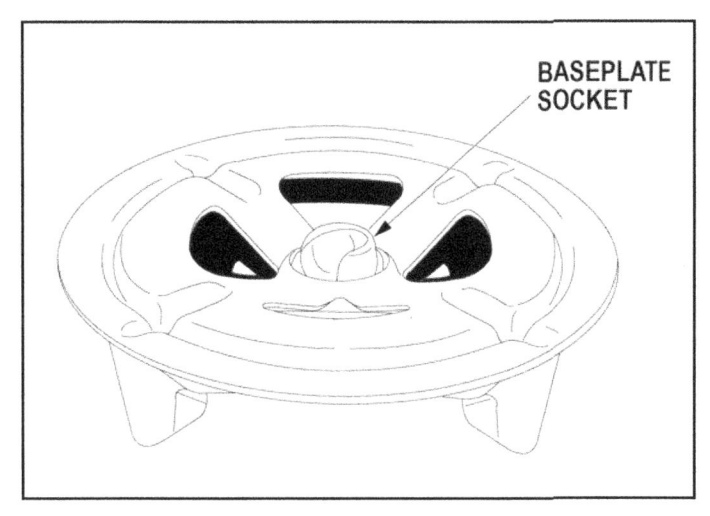

Figure 4-5. M3A1/A2 baseplate

OPERATION

4-15. This section contains information on how to prepare the M252/M252A1 81-mm mortar for firing, how to conduct safety checks, and what actions to apply to remove the cartridge from the cannon if a misfire should occur during firing.

4-16. Before the mortar is mounted, the squad must perform **premount checks**. Each squad member should be able to perform all of the following premount checks.

GUNNER

4-17. The gunner checks the baseplate and ensures that the—
- Rotating socket is free to move in a complete circle.
- Braces have no breaks, cracks, or dents M3A1/A2.
- Retaining ring is correctly located, securing the rotating socket to the baseplate.

AMMUNITION BEARER

4-18. The ammunition bearer checks the cannon and ensures that the—
- Cannon is clean and free from grease and oil, both inside and out.
- Breech plug (M253) is screwed tightly to the cannon.
- Firing pin is secured correctly.
- Blast attenuator device is secured correctly (M253).

ASSISTANT GUNNER

4-19. The assistant gunner checks the bipod and ensures that the—
- Barrel clamps are clean and dry.
- Cannon carrier is centered.

Chapter 4

- Securing strap (M177) is correctly located, securing the barrel clamps and buffers to the plain leg (M177).
- Lock release lever (leg-locking handwheel M177) is hand-tight.
- Four inches of elevation shaft are exposed, and the shaft is not bent.

SQUAD LEADER

4-20. The squad leader supervises the squad and is responsible for arranging the equipment as shown in figure 4-6.

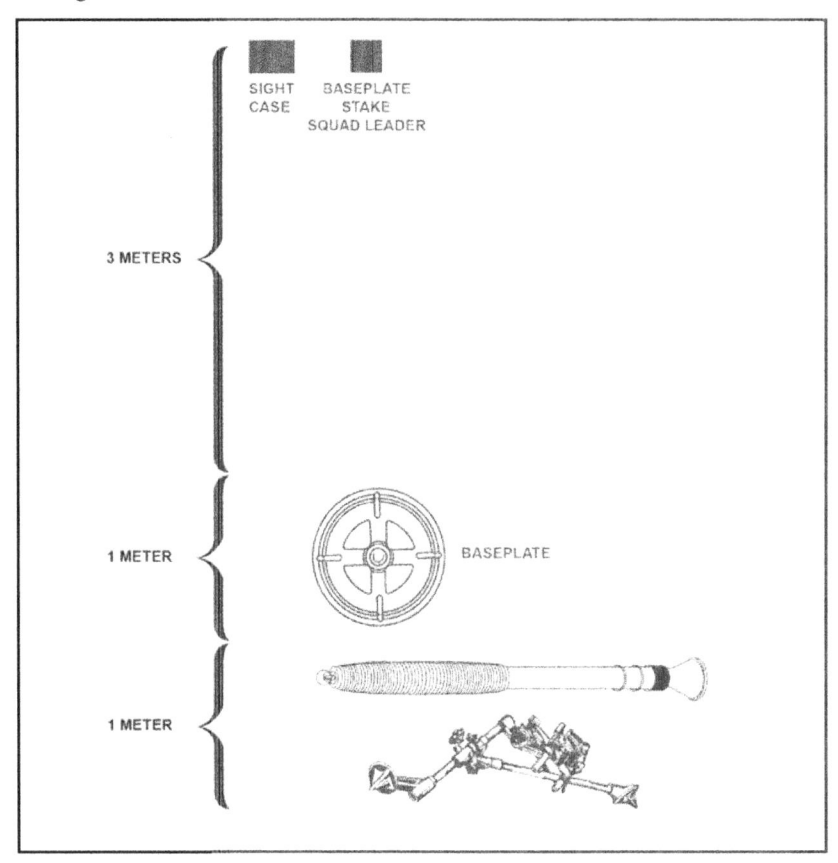

Figure 4-6. Arrangement of equipment

MOUNTING OF THE MORTAR

4-21. The squad adheres to the following procedures to mount the mortar:
- The **squad leader** picks up the sight case and the two aiming posts and moves to the exact position where the mortar is to be mounted, placing the sight case and aiming posts to the left front of the mortar position. The squad leader points to the exact spot where the mortar is to be

mounted and indicates the initial direction of fire by pointing in that direction and commanding, "ACTION."

- The **gunner** places the outer edge of the baseplate against the baseplate stake so that the left edge of the cutaway portion of the baseplate is aligned with the right edge of the stake. (See figure 4-7.) The gunner rotates the socket so that the open end is pointing in the direction of fire. During training, the gunner may use the driving stake from the aiming post case.

Note. The squad leader indicates the direction of fire when mounting.

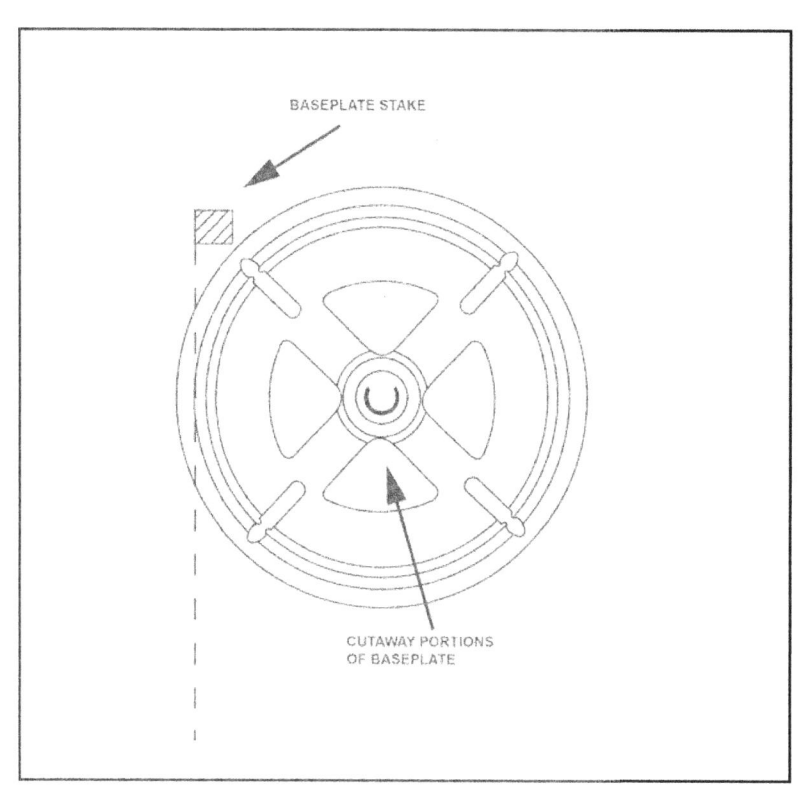

Figure 4-7. Baseplate placed against baseplate stake

- When the baseplate is in position, the **ammunition bearer** lowers the breech plug into the rotating socket and rotates the cannon a quarter of a turn to lock it. He ensures that the firing pin recess is facing upward. The ammunition bearer stands to the rear of the baseplate and supports the cannon until the bipod is fitted.

- The **assistant gunner** lifts the bipod and stands it on its legs so that the elevating hand crank (handwheel for the M177) is to the rear and the legs are to the front. He releases the cross leveling block clamp (securing strap, loosens the leg-locking handwheel and lowers the plain leg [M177]) until the locating catch engages in the recess. M177: the leg-locking handwheel must be tightened by hand, ensuring the teeth on either side are meshed correctly.

Chapter 4

- The **assistant gunner** exposes eight inches (200 millimeters) of elevation shaft, leaving the elevation hand crank (handwheel, M177) unfolded. He opens the cross-level hand crank (handwheel M177), traversing hand crank (handwheel, M177), and clamping catch (cannon clamp, M177).
- The **assistant gunner** carries the bipod to the front of the cannon and places the bipod feet on the ground 12 to 15 inches in front of the baseplate and astride the line of fire. He positions the lower clamping catch (barrel clamp, M177) against the lower stop band on the cannon and secures the clamping catch (barrel clamp, M177). The assistant gunner must ensure that the ball-shaped end of the locking rod is secured in its recess by the locking latch.
- The **gunner** removes the sight from its case, mounts it on the mortar, and sets a deflection of 3200 mils and an elevation of 1100 mils. He levels all bubbles.

SAFETY CHECKS BEFORE FIRING

4-22. The safety checks described below must be enforced before firing the mortar.

4-23. The **gunner** ensures that the—
- Cannon is locked to the baseplate, and the open end of the socket points in the direction of fire.
- Firing pin recess faces upwards.
- Bipod locking latch is locked, securing the cannon clamps.

Note. Refer to TM 9-1015-249-10 for more information on loading and firing.

- Leg-locking handwheel is hand tight.
- Check for Mask and overhead clearance.

WARNING

Keep feet clear of the mortar baseplate, as a rapidly seating baseplate can cause serious injury.

Note. Until the baseplate is seated firmly, remove the sight from the mortar before firing each round.

MASK AND OVERHEAD CLEARANCE

4-24. Since the mortar is normally mounted in defilade, there could be a mask such as a hill, tree, building, or rise in the ground. Overhead interference can be branches of trees or roofs of buildings. In any case, the gunner must ensure that the cartridge does not strike an obstacle.

4-25. In selecting the exact mortar position, the leader looks quickly for mask clearance and overhead interference. After the mortar is mounted, the gunner makes a thorough check.

4-26. The **gunner** determines mask and overhead clearance by sighting along the cannon with his eye near the breech plug. If the line of sight clears the mask, it is safe to fire. If not, the gunner may still fire at the desired range by selecting a charge zone having a higher elevation. When firing under the control of an FDC, the gunner reports that mask clearance cannot be obtained at a certain elevation.

4-27. Firing is slowed if mask clearance must be checked before each firing, but this can be eliminated if minimum mask clearance is determined. This is accomplished by depressing the cannon until the highest section of the mask is clear. The gunner levels the elevation bubble by turning the elevation micrometer knob and reading the setting on the elevation scale and elevation micrometer. This setting is the minimum mask clearance. The **squad leader** notifies the FDC of the minimum mask and overhead clearance elevation. Any target that requires an elevation outside of the determined clearance cannot be engaged from that position.

4-28. If the mask is not regular throughout the sector of fire, the **gunner** determines the minimum mask clearance as previously described. Placing the mortar in position at night does not relieve the gunner of the responsibility of checking for mask clearance and overhead interference.

4-29. The **assistant gunner** cleans the bore and swabs it dry. The **ammunition bearer** ensures that each cartridge is clean, the safety pin is present, and the ignition cartridge is in good condition.

SMALL DEFLECTION AND ELEVATION CHANGES

4-30. With the mortar mounted and the sight installed, the **gunner** makes small deflection and elevation changes by taking the following actions:
- The gunner lays the sight on the two aiming posts (placed out 50 and 100 meters from the mortar, or METT-TC dependent) on a referred deflection of 2800 mils and an elevation of 1100 mils. The mortar is within two turns of center of traverse. The vertical cross hair of the sight is on the left edge of the aiming point.
- The gunner is given a deflection change in a fire command between 20 and 60 mils inclusive. The elevation change announced must be less than 90 mils and more than 35 mils.
- As soon as the sight data are announced, the gunner places it on the sight, lays the mortar for elevation, and traverses onto the aiming post by turning the traversing handwheel and adjusting nut in the same direction. A one-quarter turn on the cross-leveling handwheel equals one turn of the traversing handwheel. When the gunner is satisfied with the sight picture, he announces, "UP."

Note. All elements given in the fire command are repeated by the squad.

- After the gunner has announced, "UP," the squad leader checks the mortar to determine if the squad drill was performed correctly.

LARGE DEFLECTION AND ELEVATION CHANGES

4-31. With the mortar mounted and the sight installed, the squad makes large deflection and elevation changes by taking the following actions:
- The **gunner** lays the sight on the two aiming posts (placed out 50 and 100 meters from the mortar) on a referred deflection of 2800 mils and an elevation of 1100 mils.

- The gunner is given a deflection and elevation change in a fire command causing the gunner to shift the mortar between 200 and 300 mils for deflection and between 100 and 200 mils for elevation.
- As soon as the sight data is announced, the gunner places it on the sight. The gunner exposes eight inches (200 millimeters) of elevation shaft and centers the buffer carrier. This ensures a maximum traversing and elevating capability after making the movement.
- The **assistant gunner** moves into position to the front of the bipod on his right knee and grasps the bipod legs (palms up), lifting until they clear the ground enough to permit lateral movement. The gunner moves the mortar as the assistant gunner steadies it. The assistant gunner tries to maintain the traversing mechanism on a horizontal plane. The gunner moves the mortar by placing his right hand over the clamp handle assembly and the left hand on the bipod leg. The assistant gunner manipulates the bipod in the direction desired by the gunner until the vertical cross hair is within 20 mils of the aiming posts. When the approximate alignment is completed, the gunner signals the assistant gunner to lower the bipod by pushing down on the mortar.
- The gunner rough-levels the cross-level bubble by making the bubble float from side to side. Then, he checks the sight picture. If he is not within 20 mils of a proper sight picture, the gunner and assistant gunner must make another large shift before continuing.
- The gunner centers the elevation bubble. He lays for deflection, taking the proper sight picture. The mortar should be within two turns of center of traverse when the task is complete.
- The open end of the socket must continue to point in the direction of fire. Normally, it can be moved by hand, although this may be difficult to do if the mortar is moved through a large arc. If required, the gunner/assistant gunner lowers the cannon so that the breech plug engages with the open end of the socket, and he uses the cannon as a lever to move the socket.
- The barrel clamps can be moved along the cannon to counter large changes in elevation, which may preclude moving the bipod. This is especially useful if the baseplate sinks deep into the ground during prolonged firing. Upon completion of any bipod movement on the cannon, the gunner ensures that the firing pin recess is facing upward.
- On uneven ground without a level surface for the bipod, the gunner can adjust the plain leg. While the assistant gunner supports the cannon, the gunner slackens the leg-locking handwheel, releases the locating catch, and positions the plain leg. Then the leg-locking handwheel must be hand tightened, ensuring the teeth are meshed correctly.

REFERRAL OF SIGHTS AND REALIGNMENT OF AIMING POSTS

4-32. Referring and realigning aiming posts ensures that all mortars are set on the same data. The section leader, acting as FDC, has one deflection instead of two or more, and takes the following actions and considerations:
- The mortar is mounted and the sight is installed. The sight is laid on two aiming posts (placed out 50 and 100 meters from the mortar, or METT-TC dependent) on a referred deflection of 2800 mils and an elevation of 1100 mils. The mortar is within two turns of center of traverse. The gunner is given an administrative command to lay the mortar on a deflection of 2860 or 2740 mils. Then the mortar is relaid on the aiming posts using the traversing crank.
- The gunner is given a deflection change between 5 and 25 mils, either increasing or decreasing from the last stated deflection, and the command to refer and realign aiming posts.

> **EXAMPLE**
>
> An example of referring and realigning aiming posts begins with the command, "REFER DEFLECTION TWO EIGHT SEVEN FIVE, REALIGN AIMING POSTS."
>
> Upon receiving the command "REFER, REALIGN AIMING POSTS," the mortar squad performs two simultaneous actions. The gunner places the announced deflection on the sight (without disturbing the lay of the weapon) and looks through the sight unit. At the same time, the ammunition bearer realigns the aiming posts. He knocks down the near aiming post and proceeds to the far aiming post. Following the arm-and-hand signals of the gunner (who is looking through the sight unit), he moves the far aiming post so that the gunner obtains an aligned sight picture. Then, he performs the same procedure to align the near aiming post.

Note: For information on malfunctions, refer to paragraphs 3-35 to 3-39 in this publication.

LOADING AND FIRING THE 81-MM MORTAR

4-33. There are procedures for loading and firing an 81-mm mortar. This includes:
- The FDC issues a fire command to the squad leader.
- The squad leader records and issues the fire command to the squad.
- The squad repeats the fire command.
- The ammunition bearer prepares the round in accordance with the fire command.
- The squad leader inspects the round before it is passed to the assistant gunner. The ammunition bearer holds the round with both hands (palms up) near each end of the round body (not on the fuze or the charges).

Note. Communication among the mortar squad is imperative in order to mitigate the probability of human error when selecting the appropriate round type, number of charges, and fuze setting. Whenever possible, the squad leader physically inspects each round prior to firing. If the situation does not facilitate physical inspection, the squad leader must visually inspect the round and verbally confirm and acknowledge with the assistant gunner that the round or rounds are all correct. The gunner remains as the last control measure to verify that all data for the fire mission is without error (deflection and elevation, sight picture, and round).

- The assistant gunner checks the round for correct charges, fuze tightness, and fuze setting.
- When both the gun and the round(s) have been determined safe and ready to fire, the squad leader gives the following command to the FDC: "NUMBER (number of mortar) GUN, UP."
- The ammunition bearer holds the round with the fuze pointed to his left. By pivoting his body to the left, the assistant gunner accepts the round from the ammunition bearer with his right hand under the round and his left hand on top of the round.

Chapter 4

- Once the assistant gunner has the round, he keeps two hands on it until it is fired.

> **CAUTION**
>
> The assistant gunner is the only member of the mortar squad who loads and fires the round.
>
> The squad leader commands, "HANG IT, FIRE," according to the method of fire given by the FDC.

> **WARNING**
>
> **Keep feet clear of the mortar baseplate, as a rapidly seating baseplate can cause serious injury.**

Note. Until the baseplate is firmly seated, remove the sight from the mortar before firing each round.

- The assistant gunner holds the round in front of the muzzle at about the same angle as the cannon. The gunner conducts final verification that the appropriate charge is set (on rounds that have been previously handled and repackaged, the potential for charges to fall off exist, especially when being handled in the limited confines of a vehicle). At the command, "HANG IT," the assistant gunner guides the round into the cannon (tail end first) to a point beyond the narrow portion of the body (about three-quarters of the round) being careful not to hit the primer or charges or disturb the lay of the mortar.
- Once the round is inserted into the cannon the proper distance, the assistant gunner shouts, "NUMBER (number of mortar) GUN, HANGING."
- At the command, "FIRE," the assistant gunner releases the round by pulling both hands down and away from the outside of the cannon. The assistant gunner ensures that he does not take his hands across the muzzle of the cannon as he drops the round.
- Once the round is released, the gunner and assistant gunner pivot their bodies so that they are both facing away from the blast.
- The assistant gunner pivots to his left and down toward the ammunition bearer, ready to accept the next round to be fired (unless major movements of the bipod require him to lift and clear the bipod off the ground).
- Subsequent rounds are fired based on the FDC fire commands.
- The assistant gunner ensures the round has fired safely before attempting to load the next round.
- The assistant gunner does not shove or push the round down the cannon. The round slides down the cannon under its own weight, strikes the firing pin, ignites, and fires.
- To avoid muzzle blast, all members of the mortar squad keep their upper body below the muzzle until the round fires.

- During an FFE, the gunner tries to level all bubbles between each round ensuring his upper body is away from the mortar and below the muzzle when the assistant gunner announces, "HANGING," for each round fired.

- The assistant gunner informs the squad leader when all rounds for the fire mission are expended, and the squad leader informs the FDC when all of the rounds are completed. For example, "NUMBER TWO GUN, ALL ROUNDS COMPLETE."

Note: For information on malfunctions, refer to paragraphs 3-35 to 3-39 in this publication.

DISMOUNTING THE MORTAR

4-34. To dismount the mortar, the squad leader commands, "OUT OF ACTION." At this command, the squad proceeds as follows:

- The gunner removes the sight and places it in the case (use deflection and elevation from the TM for storage).

- The ammunition bearer holds the cannon until the assistant gunner removes the mount. Then, the ammunition bearer rotates the cannon a quarter of a turn to unlock it from the socket and places it in an area designated by the squad leader, and retrieves the aiming posts.

- The assistant gunner disengages the cannon clamps and moves the bipod from the immediate area of the mortar position. The cannon clamps are then closed.

- With the clamps facing away from him, the assistant gunner traverses the buffer carrier to the traversing handwheel and folds the handle. He exposes one inch (25 millimeters) of the cross-level shaft and folds the handle, then exposes four inches (100 millimeters) of the elevation shaft and folds the handle. Finally, the assistant gunner loosens the leg-locking handwheel, presses the spring-loaded locating catch, and raises the plain leg behind the buffer cylinders until it touches the traversing handwheel.

- The assistant gunner tightens the leg-locking handwheel (ensuring the teeth are meshed correctly) and fastens the securing strap over the arm and around the buffers.

- The gunner returns the baseplate to the area designated by the squad leader.

4-35. The squad leader picks up the aiming posts and sight. At the command, "MARCH ORDER," the squad secures the mortar, equipment, and ammunition according to the unit SOP.

This page intentionally left blank.

Chapter 5

120-mm Mortars

The 120-mm mortar provides close-in and continuous indirect fire support to maneuver forces. The 120-mm mortar has increased range and lethality over the 81-mm and 60-mm mortars, but the mortar and ammunition are heavier. There are three versions of the 120-mm mortar: the trailer-mounted M120 mortar stowage kit (MSK), the vehicle-mounted M121, and the Stryker-mounted RMS6-L recoiling mortar. The 120-mm mortar are found in the Ranger, IBCT, ABCT and SBCT formations. This chapter discusses the organization, capabilities, and operations of the M120 and M121 120-mm mortars.

Note. Refer to TM 9-1015-250-10, TM 9-1230-205-10, and TM 9-1015-256-13&P for detailed information on the 120-mm mortars. Refer to TM 9-2355-311-10-3-1, TM 9-2355-311-10-3-2, and TM 9-2355-311-10-3-3 for more information on RMS6-L.

ORGANIZATION AND DUTIES

5-1. A mortar squad maintains and fires a single 120-mm mortar, and each member has principal duties and responsibilities.

5-2. For the mortar section to operate effectively, each squad member must be proficient in their individual duties. Performing those duties as a team member enables the mortar squad and section to perform as a fighting team. The squad, section, and platoon leaders train and lead their units.

5-3. The mortar squad for the M120 and M121 consists of four Soldiers (see figure 5-1, page 5-2)—
- Squad leader.
- Gunner.
- Assistant gunner.
- Ammunition bearer (normally functions as driver).

5-4. The **squad leader** is positioned behind the mortar in order to command and control the squad. He supervises the emplacement, laying, and firing of the mortar, and all other squad activities, and also verifies the correct round type, fuze setting, and charge prior to firing.

5-5. The **gunner** is positioned to the left side of the mortar in order to manipulate the sight, elevating handwheel, and traversing handwheel. He places firing data on the sight and lays the mortar for deflection and elevation, and makes large deflection shifts by shifting the bipod assembly and keeping the bubbles level during firing.

5-6. The **assistant gunner** is positioned to the right of the mortar, facing the cannon and ready to load. Besides loading, he swabs the bore after every 10 rounds or after each fire mission and

Chapter 5

assists the gunner in shifting the mortar when the gunner is making large deflection changes. Depending on unit SOP, the assistant gunner may help in shifting the mortar during missions. Additionally, squad members echo all fire commands.

5-7. The **ammunition bearer** stands to the right rear of the mortar and maintains the ammunition for firing by preparing the ammunition and passing it to the assistant gunner. The assistant gunner provides local security for the mortar position and may act as the squad driver.

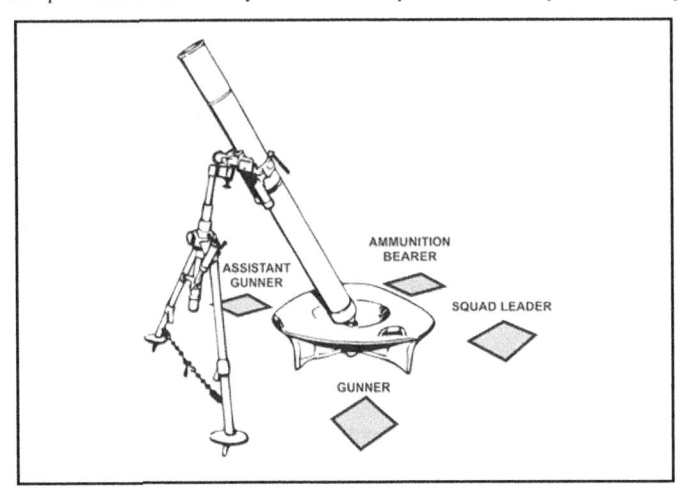

Figure 5-1. Position of squad members

COMPONENTS

5-8. This section contains the technical data and description of each component of the 120-mm mortar. (See figure 5-2, page 5-3.) The mortar is a smooth-bore, muzzle-loaded, high angle-of-fire weapon.

5-9. It consists of a cannon assembly, bipod assembly, and baseplate. The 120-mm mortar is designed to be employed in all phases and types of land warfare, and in all weather conditions.

120-mm Mortars

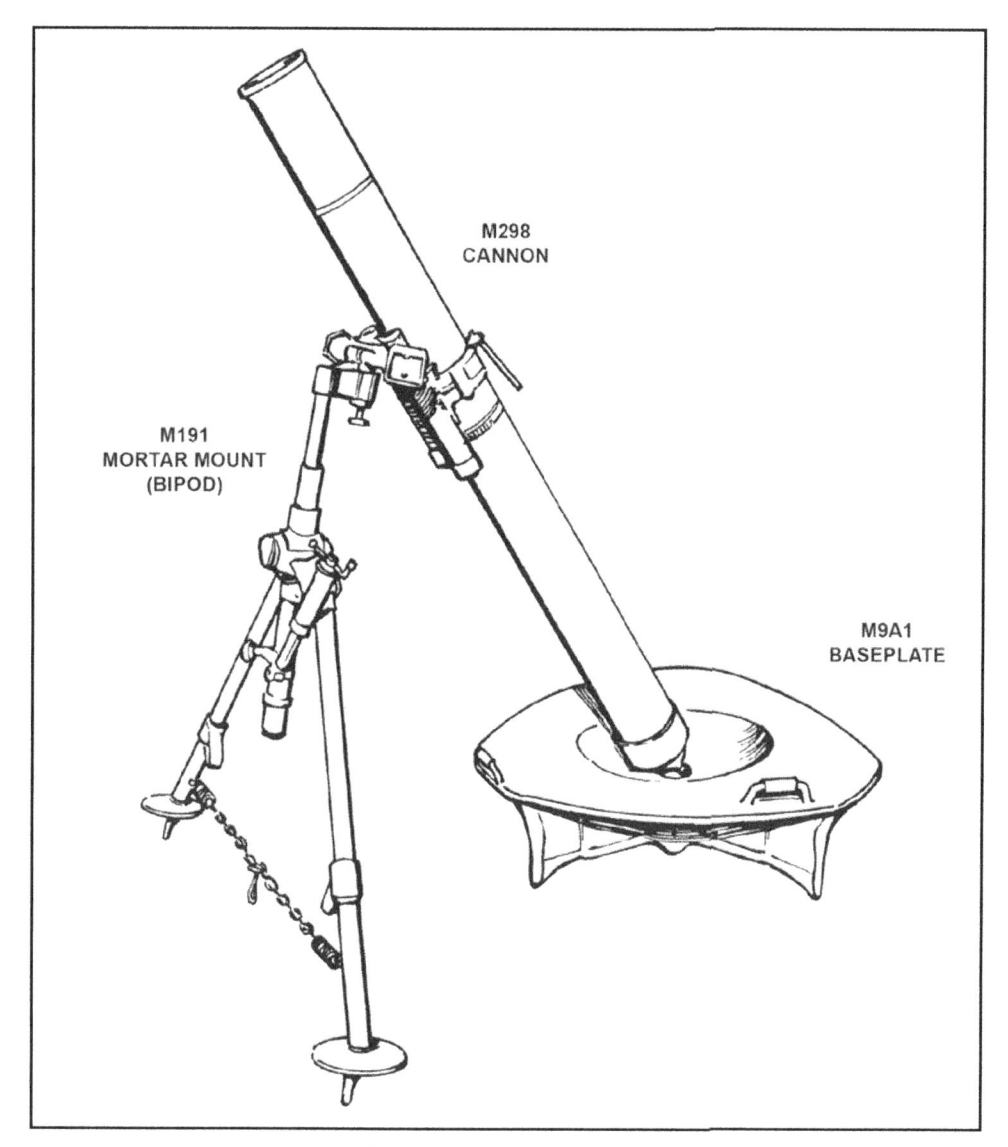

Figure 5-2. 120-mm mortar

M298 CANNON ASSEMBLY

5-10. The cannon assembly consists of two parts: the tube and the breech cap (See figure 5-3, page 5-4.) The bottom end of the tube is threaded to form a seat, which functions as a gas seal and centers the breech cap.

5-11. The breech cap screws into the base end of the tube with the front end of the breech cap mating to the seat on the tube, forming a gas-tight metal seal. The external rear portion of the

Chapter 5

breech cap is tapered and has a ball-shaped end. This end is cross-bored to help the user insert or remove the breech cap and lock it into firing position.

5-12. The breech cap houses the firing pin, which can be removed during misfire procedures as a safety measure. The squad leader must physically possess the firing pin before the mortar is considered safe.

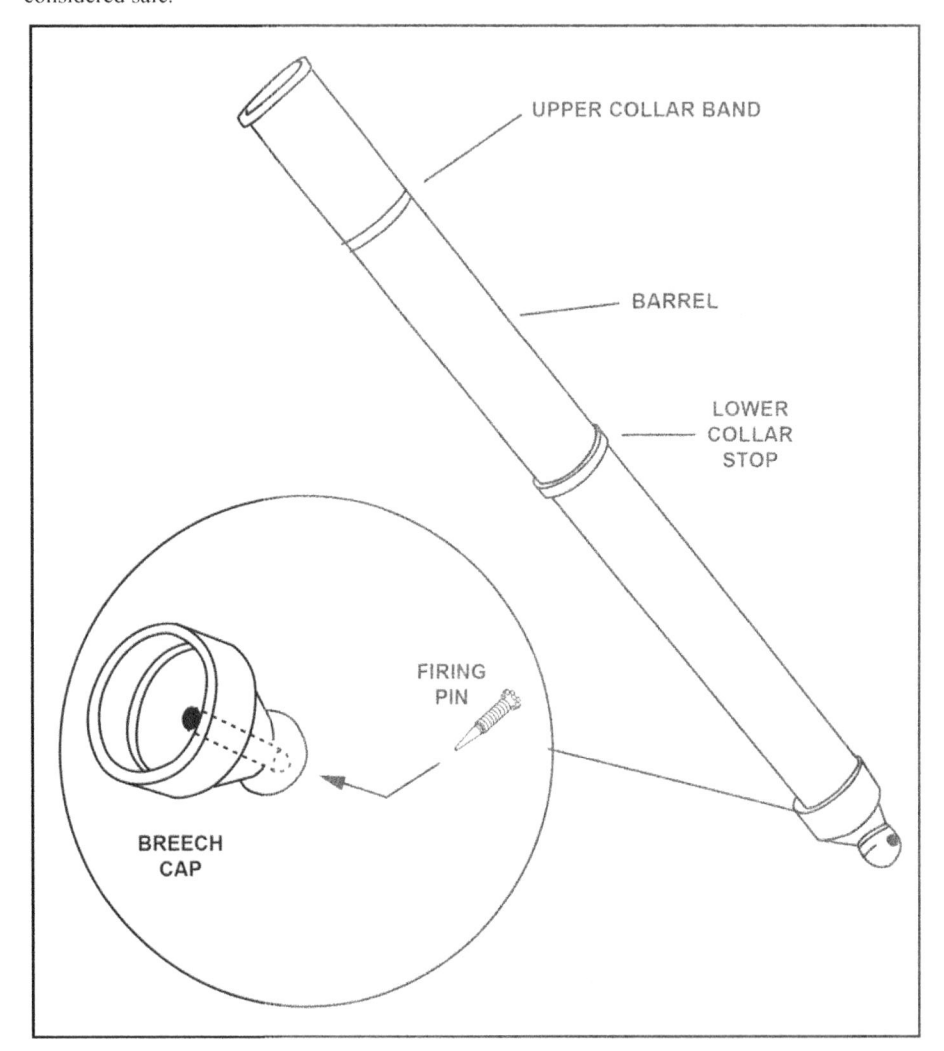

Figure 5-3. M298 cannon assembly

Note. The cannon and breech cap are serial-numbered identically. They should not be interchanged.

M191 Bipod Assembly (Carrier-/Ground-Mounted)

5-13. The M191 bipod assembly for the carrier-mounted M121 mortar (see figure 5-4) consists of the following main parts:

- Bipod leg extensions.
- Cross-leveling mechanism.
- Traversing gear assembly.
- Traversing extension assembly.
- Elevating mechanism.
- Buffer housing assembly.
- Buffer mechanism.
- Clamp handle assembly.
- Cross-leveling locking knob.
- Chain assembly.

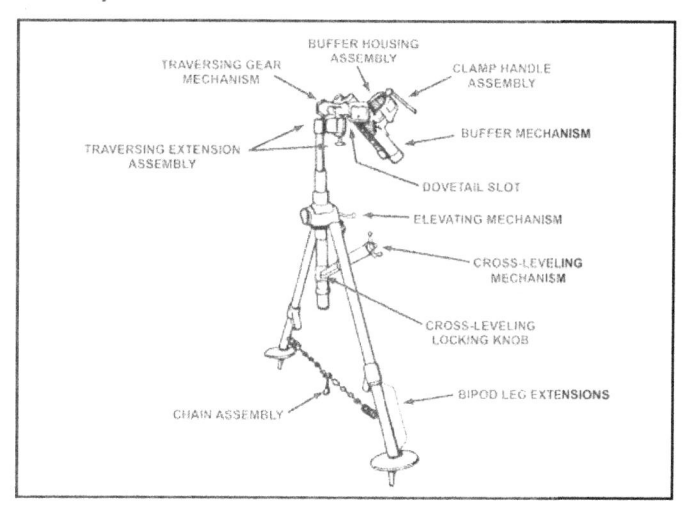

Figure 5-4. M191 bipod assembly (carrier- or ground-mounted)

M9A1 BASEPLATE

5-14. The baseplate (see figure 5-5) is shaped like a rounded triangle. It has a socket that enables a full 360-degree traverse without moving the baseplate. It has legs (spades) under the baseplate, two carrying handles, and one locking handle. M9A1 is the standard issue baseplate with the M120A1 towed mortar system.

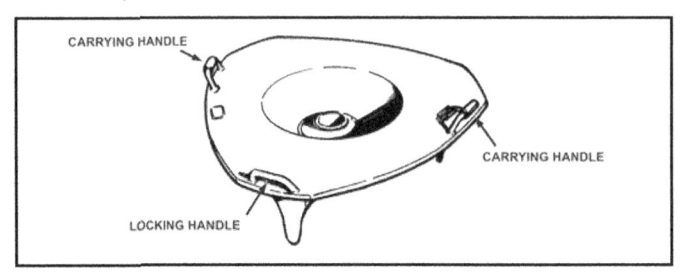

Figure 5-5. M9A1 baseplate

M120 AND M121 OPERATIONS

5-15. This section explains how to place the mortar into action by ground-mounting the weapon system; and how to conduct safety checks. Before the mortar is mounted, the squad performs **premount checks** that are described below. The premount check for the mortar cannon requires the squad to check—

- The BAD for cracks, rust, and missing parts.
- The cannon for cracks, rust, and missing, dented, or damaged parts.
- The breech assembly to ensure that it is tight.
- The white line on the breech cap to ensure that it aligns with the white line on the cannon when fully assembled.
- Around the firing pin or breech cap for evidence of gas leakage.
- The serial numbers on the barrel and breech cap to ensure that they match.
- The breech assembly for bulges, dents, or visible cracks.

Bipod Assembly

5-16. The premount check for the bipod assembly requires the squad to check—

- The bipod assembly for cracks, broken welds, or loose, missing, or damaged parts.
- To ensure the buffer housing assembly operates properly and securely holds the cannon.
- To ensure the traversing gear assembly, elevating mechanism, and cross-leveling mechanism operate smoothly and without binding through their entire range of travel.
- The buffer mechanism by pulling down on both housing tubes at the same time and ensuring that they return to their original position when released.

Mortar Baseplate

5-17. The premount check for the mortar baseplate requires the squad to check the—
- Socket for broken edges, cracks, and corrosion.
- Baseplate for cracks and broken welds.

M67 Sight Unit

5-18. The premount check for the M67 sight unit requires the squad to check the—
- Sources lit by tritium for proper illumination.
- Eyeshield for damage.
- Lenses for scratches, smears, moisture, cracks, and other obstructions.
- Reticle for clarity.
- Level vials to ensure that they are not cracked, broken, or loose in their mountings.
- Level vial covers to ensure that they are present.
- Elevation and deflection knobs to ensure that they move freely over their entire range of movement.
- Scales and index lines (all of them) to ensure that they are clear and distinct.
- Elevation and deflection knobs for movement after locking knobs are tightened. They should move no more than ± 1 mil.
- Coarse deflection scale and azimuth control dial to ensure that they rotate freely when depressed and return to their original position under spring tension when released.
- Mortar sight latch to ensure that it secures the sight to the mortar, is not loose, and has no cracks.
- Mounting surfaces to ensure that they are free of burrs and nicks.
- Radiation warning data plates to ensure that they are present and are not damaged.

PLACING A GROUND-MOUNTED 120-MM MORTAR INTO ACTION

Note. The 120-mm mortar may be transported to a location for ground operation in a number of methods including dismounted Soldiers carrying the mortar by sections, wheeled vehicle (including MSK), and carrier transported.

5-19. The mortar must be ground-mounted within one minute and 15 seconds, and meet the following conditions:
- Sight is set with a deflection of 3200 mils and an elevation of 1100 mils.
- All bubbles are centered within the outer red lines.
- Traversing extension is locked in the center position.
- Bearing is the center of traverse.
- Cannon is locked into the baseplate with the white line up (on top).
- Bipod cannon clamp is positioned and locked.
- Bipod locking knob is hand-tight.

Note. Left and right are in relation to the mortar's direction of fire.

Chapter 5

5-20. The squad uses the following procedures to place a ground-mounted 120-mm mortar into action:

- The driver/ammunition bearer exits the transport vehicle (if one was used). At the same time, the assistant gunner secures the aiming posts.
- Together, they unload the mortar and move to the firing position with the baseplate toward the direction of fire. The ammunition bearer then removes the muzzle plug.
- Once in position, the assistant gunner and ammunition bearer place the baseplate on the ground. The assistant gunner moves to the right side of the mortar to assist in mounting the mortar.
- The gunner places the sight on the left side of the mortar, then unhooks the bipod chain from the eye on the bipod leg and drops it. He then loosens the cross-leveling locking knob.

> **WARNING**
>
> Stay clear of the mortar baseplate to avoid injury from sudden release when dismounting/unloading the baseplate from a transport vehicle.

- The assistant gunner sets the bipod legs in place and spreads them until they are fully extended and the spread cable is taut.
- Once the bipod is placed on the ground, the assistant gunner tightens the cross-leveling locking knob.
- The gunner unlocks the clamp handle assembly. Now standing to the rear and straddling the cannon, the gunner grasps under the recoil buffer assembly and pulls down, sliding the cannon clamp down the cannon until it rests against the lower collar stop.
- The gunner makes sure that the white lines on the cannon and the buffer housing assembly are aligned. He retightens the clamp handle assembly until it clicks, then ensures that the firing pin is properly seated. The assistant gunner checks for slack in the spread cable and ensures that the traversing mechanism is within four turns of the center of traverse. The assistant gunner also ensures that the traversing extension is locked in the center position and that the cross-leveling locking knob is hand-tight.

Note. Part of the white line on the buffer housing assembly must overlap the white line on the cannon.

- The gunner removes the sight from the sight box and uses it to index a deflection of 3200 mils and an elevation of 1100 mils. He places the sight in the dovetail slot of the bipod. The gunner and assistant gunner then level the mortar for elevation and deflection. When the gunner is satisfied with the lay of the mortar, he announces, "GUN UP." The mortar is now mounted and ready to be laid.

Performing Safety Checks on a Ground-Mounted 120-mm Mortar

5-21. Specific safety checks must be performed before firing mortars. Most can be made visually. The gunner is responsible for physically performing the checks under the squad leader's supervision. To perform safety checks, the gunner—

- Checks for mask and overhead clearance. Then, he takes the following actions:
  - To determine mask clearance, the gunner lowers the cannon to 0800 mils elevation. He places his head near the base of the cannon and sights along the top of the cannon for obstructions through the full range of traverse.
  - To determine overhead clearance, the gunner raises the cannon to 1511 mils elevation. He places his head near the base of the cannon and sights along the top of the cannon for obstructions through the full range of traverse.

Note. If an obstruction is found at any point in the full range of traverse, both at minimum or maximum elevation, the gunner raises or lowers the cannon until the round clears the obstruction when fired. He turns the sight elevation micrometer knob until the elevation bubble is level. He reads the elevation at this point and reports the deflection and elevation to the squad leader, who reports this information to the FDC.

- Ensures the cannon is locked to the baseplate. Then, the gunner takes the following actions:
  - Locks the cannon onto the socket of the baseplate with the white line on the cannon facing up.
  - Aligns the white line on the cannon with the white line on the clamp handle assembly.
- Checks the buffer housing assembly to ensure that it is locked. This is checked by loosening the clamp handle assembly about 1/4 of a turn and retightening it until a metallic click is heard.
- Checks the cross-leveling locking knob to ensure it is hand-tight.
- Checks the spread cable to ensure that it is taut.
- Checks the firing pin to ensure that it is properly installed in the breech cap.

Performing Small Deflection and Elevation Changes on a Ground-Mounted 120-mm Mortar

5-22. The gunner receives deflection and elevation changes from the FDC in the form of a fire command. If a deflection change is required, it precedes the elevation change. To make the changes, the following actions are taken:

Notes. Small deflection changes are greater than 20 mils but less than 60 mils and small elevation changes are greater than 30 mils but less than 90 mils.

All elements of the fire command are repeated by the mortar squad.

- The gunner sets the sight for deflection and elevation. Then, takes the following actions:
  - Places the deflection on the sight by turning the deflection micrometer knob until the correct 100-mil deflection mark is indexed on the coarse deflection scale. He continues to turn the deflection micrometer knob until the remainder of the deflection is indexed on the deflection micrometer scale.

Chapter 5

- Gunner places the elevation on the sight by turning the elevation micrometer knob until the correct 100-mil elevation mark is indexed on the coarse elevation scale. He continues to turn the elevation micrometer knob until the remainder of the elevation is indexed on the micrometer scale.
- The gunner lays the mortar for deflection. Then, takes the following actions:
  - Turns the elevating hand crank to elevate or depress the mortar until the bubble in the elevation level vial starts to move. This initially rough lays the mortar for elevation.
  - Gunner looks through the sight and traverses to realign on the aiming post. He traverses half the distance to the aiming post, then cross-levels.
  - Once the vertical cross hair is near the aiming posts (about 20 mils), the gunner checks the elevation vial. He lays for elevation again by elevating or depressing the elevating mechanism. He makes final adjustments using the traversing handwheel and cross-levels by traversing half the distance and cross-leveling.
  - When the vertical cross hair is within 2 mils of the aiming posts, all bubbles are leveled, and the sight is set on a given deflection, the mortar is laid.

Note. If the given deflection exceeds left or right traverse, the gunner may choose to use the traversing extension assembly. This gives an additional 90 mils in traverse to avoid moving the bipod. To use the traversing extension, the gunner pulls down on the traversing extension locking knob and shifts the cannon left or right. Then, relocks the traversing extension locking knob by ensuring it is securely seated.

PERFORMING LARGE DEFLECTION AND ELEVATION CHANGES ON A GROUND-MOUNTED 120-MM MORTAR

5-23. The gunner receives deflection and elevation changes from the FDC in the form of a fire command. If a deflection change is required, it always precedes the elevation change. The gunner lays the mortar for large deflection and elevation changes. To make the changes, the following actions are taken:

Note. Large deflection changes are greater than 200 mils but less than 300 mils and large elevation changes are greater than 100 mils but less than 200 mils.

- The gunner receives a deflection and elevation change in the form of an initial fire command.

  Note. All elements of the fire command are repeated by the mortar squad. Preshift is allowed.

- As soon as the gunner receives the data, he places it on the sight and elevates or depresses the mortar to float the elevation bubble.
- The assistant gunner moves into position to the front of the bipod on his right knee and grasps the bipod legs (palms up), lifting until they clear the ground enough to permit lateral movement. The gunner moves the mortar as the assistant gunner steadies it. The assistant gunner tries to maintain the traversing mechanism on a horizontal plane. The gunner moves the mortar by placing his right hand over the clamp handle assembly and his left hand on the bipod leg, pushing or pulling the bipod in the direction desired until the vertical cross hair is within 20

120-mm Mortars

mils of the aiming posts. When the approximate alignment is completed, the gunner signals the assistant gunner to lower the bipod by pushing down on the mortar.

- The gunner then floats the deflection bubble and looks into the sight to see if he is within 20 mils of his aiming posts. If not, the process is repeated until within 20 mils.
- The gunner and assistant gunner level the mortar for elevation. If after leveling the mortar for elevation the vertical cross hair of the M67 sight is within 20 mils of the aiming post, he then centers the deflection bubble and takes up the proper sight picture by traversing half the distance to the aiming posts and cross-leveling.
- The assistant gunner observes the gunner traversing to ensure that he stays within four turns of center-of-traverse. Should the gunner traverse away from center-of-traverse, the assistant gunner advises and instructs the gunner to center back up. The gunner center traverses, and with the help of the assistant gunner, shifts the bipod again and repeats steps 3 through 5.

Note. If the vertical cross hair of the M67 sight is not within 20 mils of the aiming post after leveling the mortar, then steps 3 through 5 are repeated.

- The gunner makes minor adjustments as necessary to obtain a correct sight picture and does a final check of the bubbles and center-of-traverse, and announces, "UP."

Note: For information on malfunctions, refer to paragraphs 3-35 to 3-39 in this publication.

REFERRAL OF SIGHTS AND REALIGNMENT OF AIMING POSTS

5-24. Referring and realigning aiming posts ensures that all mortars are set on the same data. The FDC has one deflection instead of two or more. The squad uses the following procedures to refer the sight and realign the aiming posts:

- The mortar is mounted and the sight is installed. The sight is laid on two aiming posts (placed out 50 and 100 meters from the mortar) on a referred deflection of 2800 mils and an elevation of 1100 mils. The mortar is within two turns of center of traverse. The gunner is given an administrative command to lay the mortar on a deflection of 2860 or 2740 mils. Then the mortar is relaid on the aiming posts using the traversing crank.
- The gunner is given a deflection change between 5 and 25 mils, either increasing or decreasing from the last stated deflection, and the command to refer and realign aiming posts.

LOADING AND FIRING OF THE GROUND-MOUNTED 120-MM MORTAR

5-25. There are procedures for loading and firing a ground-mounted 120-mm mortar. This includes:

- The FDC issues a fire command to the squad leader.
- The squad leader records and issues the fire command to the squad.
- The squad repeats the fire command.
- The ammunition bearer prepares the round according to the fire command.
- The squad leader inspects the round before it is passed to the assistant gunner. The ammunition bearer holds the round with both hands (palms up) near each end of the round body (not on the fuze or the charges).

Chapter 5

> *Note.* Communication among the mortar squad is imperative to mitigate the probability of human error when selecting the appropriate round type, number of charges, and fuze setting. Whenever possible, the squad leader physically inspects each round before firing. If the situation does not facilitate physical inspection, the squad leader must visually inspect the round and verbally confirm and acknowledge with the assistant gunner that the round or rounds are correct. The gunner is the last control measure to verify that all data for the fire mission is without error (deflection and elevation, sight picture, and round).

- The assistant gunner checks the round for correct charges, fuze tightness, and fuze setting.
- When both the gun and the round(s) have been determined safe and ready to fire, the squad leader then gives the following command to the FDC, "NUMBER (number of mortar) GUN, UP."
- The ammunition bearer holds the round with the fuze pointed to his left. By pivoting his body to the left, the assistant gunner accepts the round from the ammunition bearer with his right hand under the round and his left hand on top of the round.
- Once the assistant gunner has the round and keeps two hands on it until it is fired.

CAUTION

The assistant gunner is the only member of the mortar squad who loads and fires the round.

The squad leader commands, "HANG IT, FIRE," according to the method of fire given by the FDC.

WARNING

Keep feet clear of the mortar baseplate, as a rapidly seating baseplate can cause serious injury.

> *Note.* Until the baseplate is seated firmly, remove the sight from the mortar before firing each round.

- The assistant gunner holds the round in front of the muzzle at about the same angle as the cannon. The gunner conducts final verification that the appropriate charge is set (on rounds that have been previously handled and repackaged, the potential for charges to fall off exist, especially when being handled in the limited confines of a vehicle). At the command, "HANG IT," the assistant gunner guides the round into the cannon (tail end first) to a point beyond the narrow portion of the body (about three-quarters of the round) being careful not to hit the primer or charges or disturb the lay of the mortar.
- Once the round is inserted into the cannon the proper distance and the assistant gunner shouts, "NUMBER (number of mortar) GUN, HANGING."

120-mm Mortars

- At the command, "FIRE," the assistant gunner releases the round by pulling both hands down and away from the outside of the cannon. The assistant gunner ensures that he does not take his hands across the muzzle of the cannon while dropping the round.
- Once the round is released, the gunner and assistant gunner take a full step toward the rear of the weapon while pivoting their bodies so that they are both facing away from the blast.
- The assistant gunner pivots to his left and down toward the ammunition bearer, ready to accept the next round to be fired (unless major movements of the bipod require him to lift and clear the bipod off the ground).
- Subsequent rounds are fired based on the FDC fire commands.
- The assistant gunner ensures the round has fired safely before attempting to load the next round.
- The assistant gunner does not shove or push the round down the cannon. The round slides down the cannon under its own weight, strikes the firing pin, ignites, and fires.
- The assistant gunner and gunner, as well as the remainder of the mortar squad, keep their upper body below the muzzle until the round fires to avoid muzzle blast.
- During an FFE, the gunner tries to level all bubbles between each round ensuring his upper body is away from the mortar and below the muzzle when the assistant gunner announces, "HANGING," for each round fired.
- The assistant gunner informs the squad leader when all rounds for the fire mission are expended, and the squad leader informs the FDC when all of the rounds are completed. For example, "NUMBER TWO GUN, ALL ROUNDS COMPLETE."

TAKING THE 120-MM MORTAR OUT OF ACTION

5-26. To take the 120-mm mortar out of action, the squad leader commands, "OUT OF ACTION." Then, each member of the squad does the following:

- The ammunition bearer retrieves the aiming posts, places them in their case, and puts the case on the right side of the mortar.
- The gunner places the deflection and elevation from TM for storage.
- The gunner removes the sight from the dovetail slot and places it in the sight case. Then places the sight mount cover back on and secures it with the snap button.
- The gunner loosens the clamp handle assembly.
- The gunner slides the buffer housing assembly up the cannon until the white line on the clamp handle assembly is aligned with the white line on the cannon near the muzzle. The gunner tightens the buffer housing assembly until he hears a metallic click, and continues turning the handle until it is parallel to the cannon.
- The gunner centers the traversing extension and the traversing mechanism.
- The gunner lowers the elevation so that about four fingers (three to four inches) of the elevating mechanism remains exposed.
- The assistant gunner loosens the cross-leveling locking knob, while the gunner steadies the cannon.

Note. The mortar squad prepares the mortar system for carrying, mounting, loading, or storage; based on the method used for transportation.

- The ammunition bearer emplaces the muzzle cap.
- The gunner secures the sight case.

Chapter 5

- The assistant gunner secures the aiming post case.
- The squad leader ensures that all equipment is accounted for and secured properly.
- The squad leader announces, "NUMBER TWO GUN, UP."

M326 120-MM MORTAR STOWAGE KIT OPERATIONS

5-27. The M326 MSK is a powered device that allows a 120-mm mortar to be carried as a single unit on an M1101 trailer. The device uses a mortar support strut to hold the mortar tube, base plate, and bipod solidly together. This assembly is emplaced or recovered by a hydraulic driven winch. A manual lift winch and strap also are available for use as a manual backup.

Note. Refer to most updated TMs for detailed information on the M326, M150 or M1101.

5-28. The MSK eliminates the need for the mortar to be disassembled for deployment and retrieval. The result is improved emplacement and displacement times while reducing squad effort, and minimizing damage to equipment due to the hazards and wear associated with assembly, disassembly, storage, and transport.

5-29. Once the 120-mm mortar is emplaced, the support strut is disengaged and the hauling platform is moved before mortar firing begins. The M326 is provided with a standard North Atlantic Treaty Organization (NATO) cable interface. This interface allows for charging of the M326 batteries using the prime mover.

5-30. The M326 120-mm mortar stowage kit (see figure 5-6, page 5-15) has eight major components—

- Lift drive assembly.
- Power box assembly.
- Tube support assembly.
- Junction box assembly.
- Control pendant assembly.
- Mounting subframe.
- Guide rail assemblies.
- Support strut assembly.

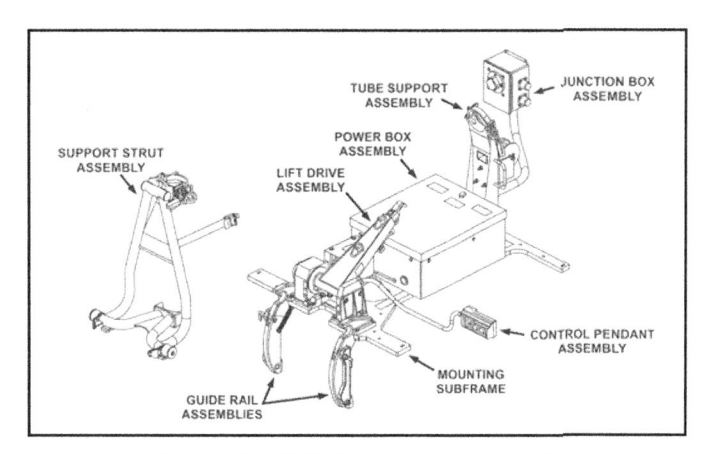

Figure 5-6. M326 mortar stowage kit

BASIC ISSUE ITEMS FOR THE M326 MORTAR STOWAGE KIT

5-31. The basic issue items box holds the—
- Extension chain assembly.
- Ammunition rack cover (tilted), two each.
- Ammunition rack cover (vertical), one each.
- Funnel.
- NATO slave cable, 20 feet.
- 120-mm mortar muzzle plug.
- Screwdrivers, one flat tip and one cross tip.
- Adjustable wrench, two each.

5-32. There are three ammunition racks mounted on the M1101 trailer (see figure 5-7, page 5-16). Two of the ammunition racks are mounted on either side of the M1101 trailer with the third rack mounted to the front of the trailer. Each rack holds eight rounds, in either cardboard or mono packs. The side racks are mounted at a 10 degree angle for easy access from outside of the trailer.

5-33. The rack mounted to the front of the trailer is used primarily for ammunition requiring vertical storage. The holes in the design are for aeration and drainage. Each ammunition rack consists of the rack, eight ammunition retainer caps, and two retainer straps. (See figure 5-8, page 5-16.) The ammunition rack support bar is used to provide support for the two tilted ammunition racks.

Chapter 5

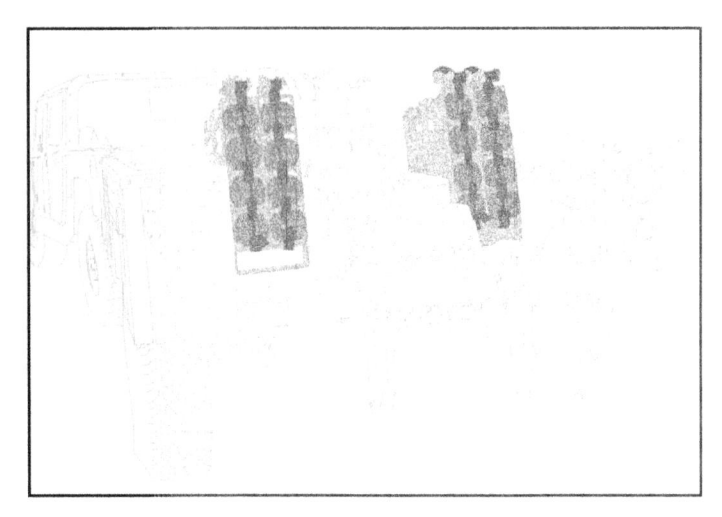

Figure 5-7. Ammunition racks mounted on the M1101 trailer

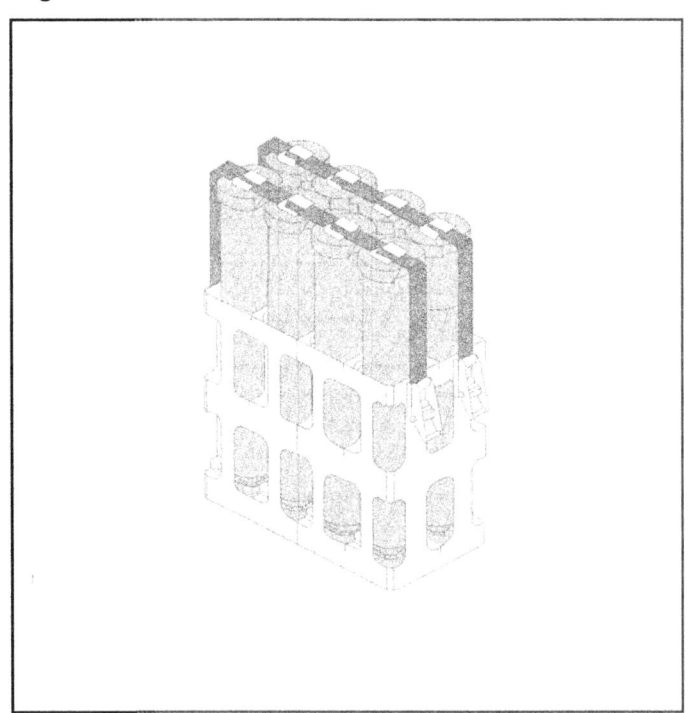

Figure 5-8. Ammunition racks unmounted

5-34. The mortar squad consists of four members: the squad leader, gunner, assistant gunner, and ammunition bearer/driver. Units assign personal and seating assignments based on action drills, mission essential task list, and unit SOPs. Basic seating assignments are shown in figure 5-9.

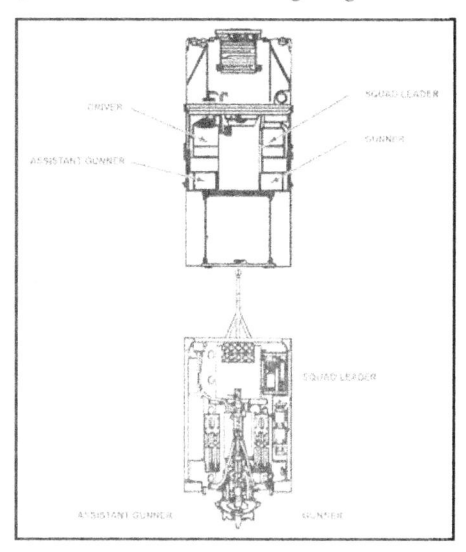

Figure 5-9. Squad member positions

WARNING

When the M1101 trailer has been disconnected from the HMMWV, the rear supports legs of the trailer must be extended to prevent the trailer from tipping. If the squad attempts to deploy or retrieve an M120 mortar the HMMWV must be disconnected from the trailer.

Chapter 5

> **WARNING**
>
> Ensure that when the M1101 is disconnected from the HMMWV vehicle with M326 120-mm mortar stowage kit hardware, that the rear supports are extended to prevent the trailer from tipping due to an unbalanced load if the squad attempts to retrieve or deploy an M120-mm mortar. Ensure all personnel are clear of the path of moving hardware and platform when performing powered displacement or emplacement operations. Failure to heed warnings could result in crush hazards during operations or exposure to a high pressure leak in the event of a hardware failure. Ensure personnel remain six feet from descending or ascending mortar tube to prevent being struck by M150 mortar hardware.

> **WARNING**
>
> When operating or maintaining M326 120-mm mortar stowage kit hardware, personnel must wear gloves in order to prevent thermal burns or frostbite to hands in extreme ambient conditions (temperature is less than 32 or more than 120 degrees Fahrenheit). Personnel must wear gloves during operations or maintenance activities. Wearing gloves mitigates exposure to laceration hazards. When performing operations or maintenance, ensure adequate personnel are available for lifting and moving of equipment exceeding a single-person lift (50 pounds). Failure to do so may result in back strain injuries.

EMPLACING THE M150 120-MM GROUND-MOUNTED MORTAR USING THE M326 120-MM MORTAR STOWAGE KIT

5-35. There are several fundamental steps to emplace the M150 using the M326 120-mm mortar stowage kit, and are described in the paragraphs below.

Step 1. The ammunition bearer/driver (after placing the vehicle in PARK mode with emergency brake engaged) and assistant gunner exit the vehicle from the left side. The squad leader and gunner exit from the right side the vehicle.

Note. The following actions are conducted simultaneously.

Step 2. The squad leader disconnects the fire control computer (FCC) from the visor mount. Then the squad leader disconnects the FCC 9w1 P1 cable. The squad leader moves to the trailer with the FCC and inserts it into the electronics rack. Then plugs the 9w1 P1 cable into the FCC from the left side and continues to setup the FCC and any associated equipment in the electronic rack.

Step 3. The ammunition bearer/driver secures the trailer legs and attaches them to the trailer (if being detached from prime mover). The assistant gunner ensures that the junction box is set to direct (if the vehicle is to remain in position, if not it is set to OFF).

Step 4. The gunner moves to the trailer and releases the tube support clamp on the tube support assembly. The gunner (right) and assistant gunner (left) move to the rear of the trailer where they lower the guide ramps.

Note. To skip to manual mode steps go to paragraph 5-41.

5-36. In POWER mode—

Step 1. The assistant gunner ensures that the lift drive selector is in the POWERED position and fully seated.

Step 2. The gunner secures the control pendant and turns the power switch to the ON position to check the battery condition gauge on the power supply enclosure for sufficient power.

Note. If there is no power, take the following troubleshooting steps. Ensure the hydraulic pump unit isolator switch is ON; if not, turn it on. Check the power fuse, replace if unserviceable and notify maintenance. If the troubleshooting steps do not solve the power issue, emplace the 120-mm mortar manually.

Step 3. The gunner commands, "CLEAR." The squad members respond with "ALL CLEAR." If no hazard exists, the gunner continues to next step. The gunner presses the DOWN button on the control pendant; the assistant gunner assists in lowering the weapon system by standing to the left or right and slightly to the front of weapon. When the weapon system is halfway down, the assistant gunner guides the system to the ground by pulling on the baseplate handle until the lift arm hook disengages from the mortar strut at a minimum of six inches from the mortar strut.

Step 4. When the mortar is on the ground, the gunner raises the lift arm hook by pressing the UP button on the control pendant, returning the lift arm to its stowed position and then turns the control pendant off. The ammunition bearer/driver (left) and gunner (right) then place the guide ramps in the travel (UP) position.

Chapter 5

5-37. In MANUAL mode—

Step 1. The assistant gunner ensures that the lift drive selector is in the MANUAL position and fully seated. (See figure 5-10.)

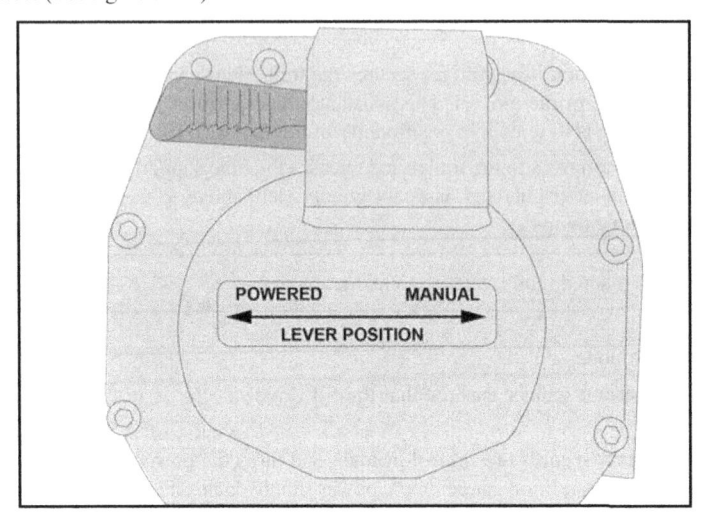

Figure 5-10. Lift drive selector

Note. If the lift drive selector cannot be moved, apply pressure on the baseplate by pushing towards the front of the trailer.

Step 2. The ammunition bearer/driver mounts the trailer and performs the following: while facing the open end of the M1101 trailer, the ammunition bearer places his left hand on the manual winch drum and turns the handle in a counter-clockwise direction. Once the handle is loose, maintain outward pressure on the handle and grasp the winch strap. Feed the strap through the guide hole. (See figure 5-11.)

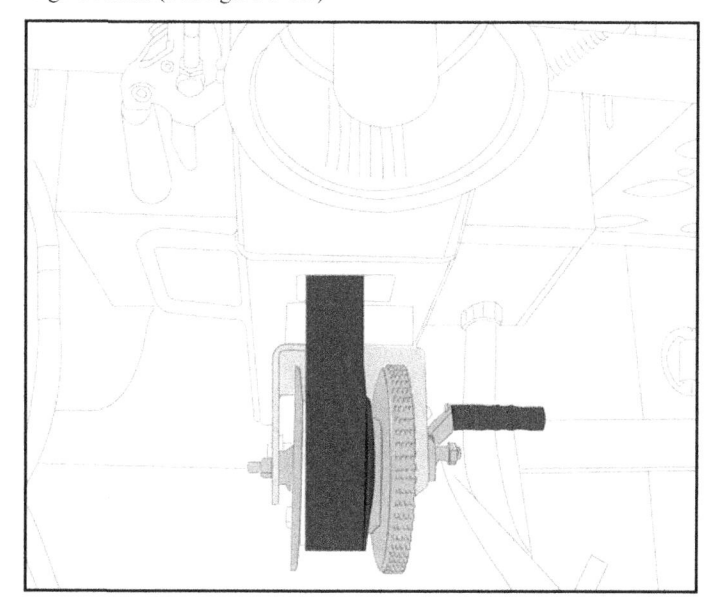

Figure 5-11. Manual winch drum and guide hole

Step 3. The gunner pulls the strap through the strap guide hole in the mortar support and attaches the strap clamp to support the strut eye hook. (See figure 5-12.) The assistant gunner releases the manual lift strap support arm from the lift arm and lets it rest on the battery box.

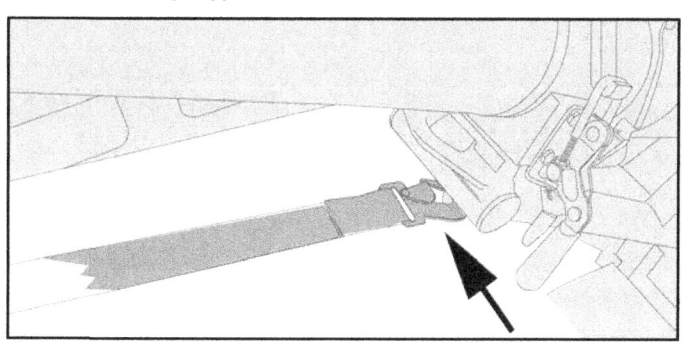

Figure 5-12. Support strut eye hook

Step 4. The ammunition bearer turns the handle in clockwise direction to remove any slack in the strap (see figure 5-11).

Chapter 5

Step 5. The ammunition bearer unlatches the barrel clamp (see figure 5-13) and turns the handle in a counterclockwise direction.

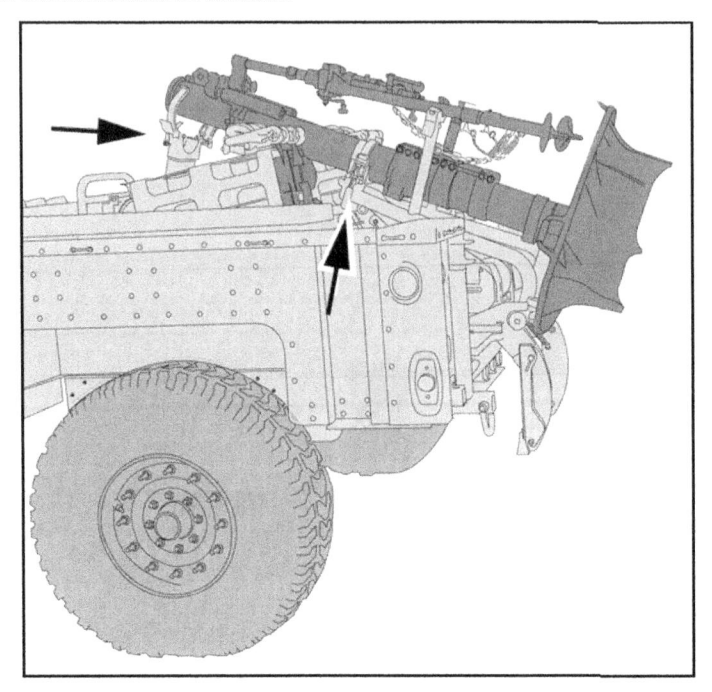

Figure 5-13. Barrel clamp

Note. Ensure the strap is taut and properly aligned with the manual winch standoff as shown in figure 5-14, page 5-23.

Step 6. The gunner grasps the baseplate as shown in figure 5-15, page 5-24, and helps guide the system to the apex.

> **CAUTION**
> The lift arm assembly is disengaged and will drop unless the strap is kept taut.

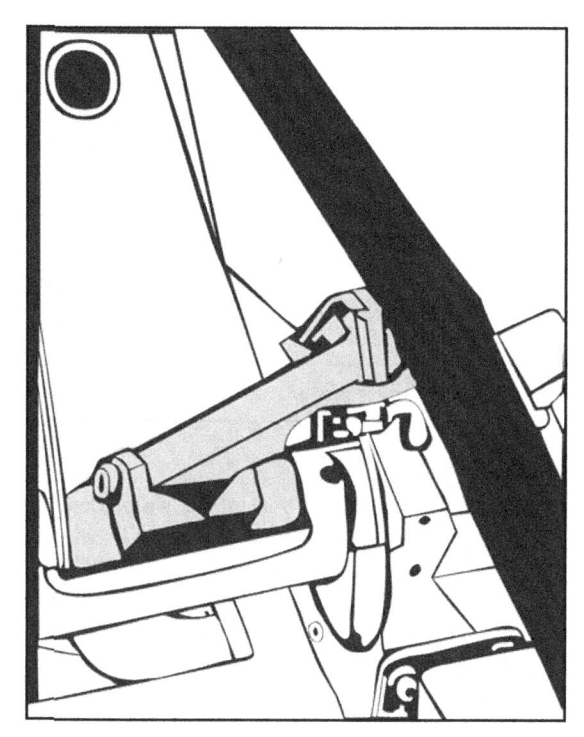

Figure 5-14. Winch strap aligned with the manual winch standoff

Chapter 5

Figure 5-15. Gunner guide to apex

Step 7. The ammunition bearer/driver continues turning the manual winch in a counter-clockwise direction until the lift hook is disengaged from the support strut assembly and rests on the ground.

Note. Ensure to move the M1101 trailer forward at least three feet.

5-38. In POWER and MANUAL modes—

Step 1. The assistant gunner moves to the front of the weapon and loosens the cross-level locking knob and unhooks the bipod chain from the eye on the bipod leg. Then, the gunner and assistant gunner disengage the bipod legs by pulling on the spring-loaded catches on the support strut bipod arms until the bipod legs are free.

Step 2. The ammunition bearer/driver stands on the baseplate and waits for the gunner to tell him to unclamp the support strut tube clamp. The gunner grasps the clamp handle assembly on the mortar and barrel by the muzzle of the weapon and commands the ammunition bearer to disengage the support strut tube clamp. The assistant gunner holds the bipod legs just above the spread chain; pulls and guides the mortar forward away from the trailer. The assistant gunner guides the bipod legs to a point (about two feet) in front of the baseplate and lowers them to the ground. At this point the mortar is free from the support strut.

Step 3. Once the weapon is on the ground, the gunner unlocks the clamp handle assembly. The gunner, who is now standing to the rear of the cannon, grasps under the recoil buffer assembly and pulls down, sliding the cannon clamp down the cannon until it rests against the lower collar stop, firing pin is facing skyward, white lines on the barrel and the buffer housing assembly are aligned, and the gunner retightens the clamp handle assembly until one metallic click is heard.

Step 4. The assistant gunner checks for slack in the spreader chain and ensures the traversing mechanism is within four turns from center of traverse. The assistant gunner also ensures the traversing extension is locked in the center position and the cross-leveling locking handle is hand tight.

Step 5. The ammunition bearer/driver secures the mortar support strut and moves the mortar support strut to a location away from the weapon system, ensuring that no cables are under the baseplate.

120-mm Mortars

> **CAUTION**
>
> The pointing device mount assembly–dynamic quick release (PDMA-DQR) is a two-man lift: It weighs 67 pounds. Failure to adequately support assembly can injure personnel or damage critical equipment.

Step 6. The assistant gunner unlocks the pointing device mount assembly-dynamic quick release (PDMA-DQR) from the stowed position on the trailer; the gunner secures the two dynamic quick release handles. The gunner and the assistant gunner lift the PDMA-DQR and carry it to the gun system. They secure the PDMA-DQR to the pointing device mount bracket. The gunner then secures the gunners display, places it onto the bipod leg, and tightens the clamp onto the ramball. If using a M120 ground-mounted mortar system not equipped with the MFCS, then the mortar squad places the mortar into action in accordance with unit SOP.

- The mortar is now mounted.

EMPLACING THE ELECTRONIC RACK

> **CAUTION**
>
> The electronic rack is a three-man lift. It weighs 144 pounds. Failure to adequately support the assembly can injure personnel or damage critical equipment.

5-39. To emplace the electronic rack, the following procedures are conducted:

- The ammunition bearer/driver mounts the trailer and loosens the tie down straps and ensures the electronics rack is not connected to the NATO adapter at the junction box. The squad leader secures one side of the upper electronics rack and the assistant gunner secures the other side of the upper electronics rack. In one fluid motion, all three lift the electronics rack up to the side wall of the trailer. The ammunition bearer/driver dismounts the trailer and helps to lower the electronics rack to the ground.

- The gunner and assistant gunner then manipulate the mortar for elevation and deflection until the gunners display reads, LAYED, then the gunner announces, "UP," and steps away from the mortar.

- The ammunition bearer/driver moves the vehicle forward approximately three feet (if directed). If the vehicle has been moved, the ammunition bearer/driver places the vehicle in PARK mode with safety brakes engaged and turns the engine off. Once the vehicle is parked the ammunition bearer/driver moves to the left rear of the trailer and waits for further commands.

DISPLACING THE M150 120-MM GROUND-MOUNTED MORTAR USING THE M326 120-MM MORTAR STOWAGE KIT

5-40. The fundamental steps to displace the M150 using the M326 are as follows:

Chapter 5

> **CAUTION**
> The winch strap is rated at 1670 pounds lift capability. Ensure the strap is properly aligned with the manual winch standoff. The strap has a breaking capacity of 2600 pounds.

- The ammunition bearer/driver (after placing the vehicle in PARK with emergency brake engaged) moves to the rear of the vehicle to assist the mortar squad in displacement procedures. If using a M120 ground-mounted mortar system, not equipped with the MFCS, then the mortar squad will take the mortar out of action according to the unit SOP.

Note. When transferring the gunner's display from firing position to stowed position, it is not necessary to disconnect cable 9W2 from the gunner's display.

- The gunner removes the gunner's display and mounts it to the RAM ball on the trailer. The gunner positions the locking clamp over the trailer RAM ball mount. Then positions the gunner's display mount assembly RAM ball in the other end of locking clamp. The gunner tightens the T-handle on locking clamp to secure the gunner's display mount assembly to the trailer RAM ball mount.

Note. Cable 9W2 plugs P4 and P5 remain connected to pointing device, and pointing device remains connected to PDMA-DQR when removing PDMA-DQR from gun tube.

- The gunner grasps the two dynamic quick release locking handles firmly, depresses the locking handle springs, and rotates locking handles up to vertical, releasing the locking handle spring. The gunner verifies locking pins have disengaged from the upper bracket. The assistant gunner helps the gunner pull the PDMA-DQR towards the muzzle of mortar tube and lifting the assembly from upper and lower mortar tube clamps, ensuring the PDMA-DQR lower support pins clear the lower barrel clamp. The assistant gunner grasps the PDMA-DQR and assists the gunner in carrying it to the trailer stowage mount. They position the PDMA-DQR over the stowage mount and align the support pins and rear bracket locking legs with the forward and rear stowage brackets.

STOWAGE MOUNT

5-41. The assistant gunner depresses the handle locking spring to disengage the locking pin from the forward stowage bracket. The gunner pushes the PDMA-DQR forward until support pins and rear locking legs have fully engaged with the forward and rear stowage bracket. Assistant gunner grasps locking handle and depresses the handle locking spring. The assistant gunner turns locking handle to the vertical position until locking pin engages with the upper locking hole in the forward stowage bracket.

120-mm Mortars

> **CAUTION**
> Locking features must be fully engaged in forward and rear stowage brackets. Failure to comply can damage locking device or allow pointing device to become loose from the stowage bracket.

LOADING THE MORTAR ONTO THE VEHICLE

5-42. To load the mortar onto the vehicle the mortar squad does the following:

- The assistant gunner moves to the front of the weapon and traverses it to within four turns of center and confirms there are four fingers of elevation shaft showing. Then the assistant gunner ensures the barrel clamp is moved to within four inches of the white line on the barrel. The assistant gunner continues by making sure the cross-level locking handle is loose. The ammunition bearer/driver secures and locks the support strut assembly onto the baseplate. (See figure 5-16A.) The ammunition bearer/driver opens the support strut tube clamp to accept the mortar tube. (See figure 5-16B, page 5-28.)

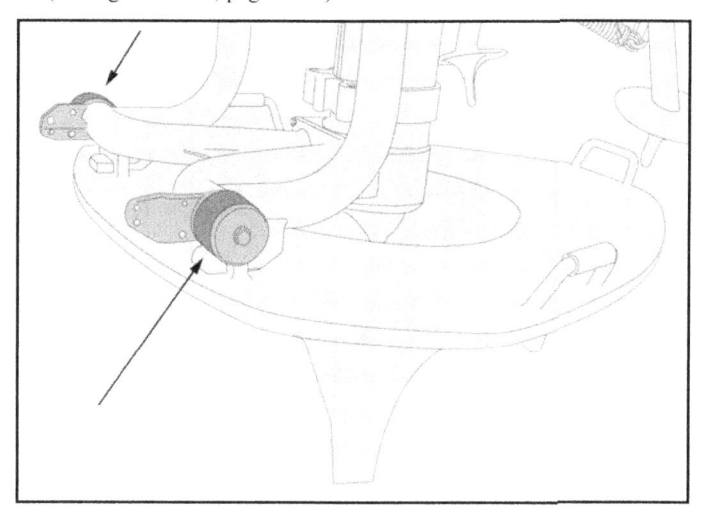

Figure 5-16A. Support strut and baseplate brackets

Chapter 5

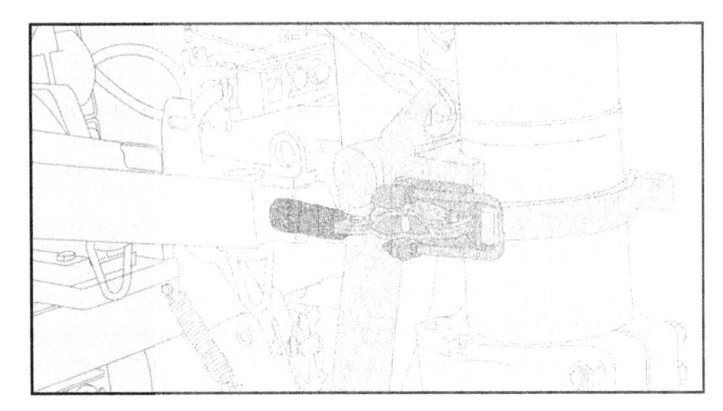

Figure 5-16B. Tube clamp

- With the aid of the gunner, the assistant gunner grasps the bipod legs and lifts the bipod until the tube is seated into the support strut clamp, and hooks the bipod chain from the eye on the bipod leg.
- Ammunition bearer/driver ensures the barrel white line is aligned with the center of the support strut tube clamp. The ammunition bearer/driver latches and closes the support strut tube clamp. Assistant gunner and gunner push the bipod legs into the bipod arm assembly. (See figure 5-17.) The mortar is ready for lifting. (See figure 5-18.)

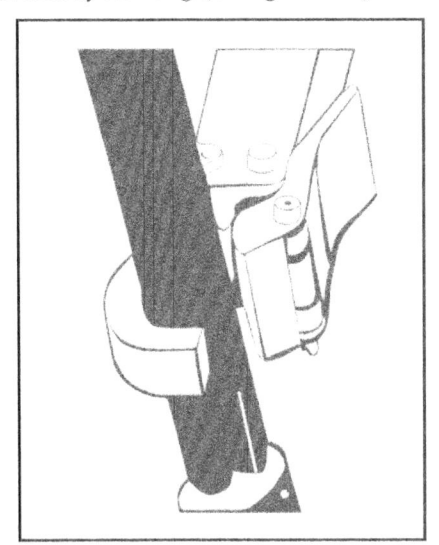

Figure 5-17. Support strut tube clamp

5-28　　　　　　　　　　　　　　TC 3-22.90　　　　　　　　　　　　　17 March 2017

120-mm Mortars

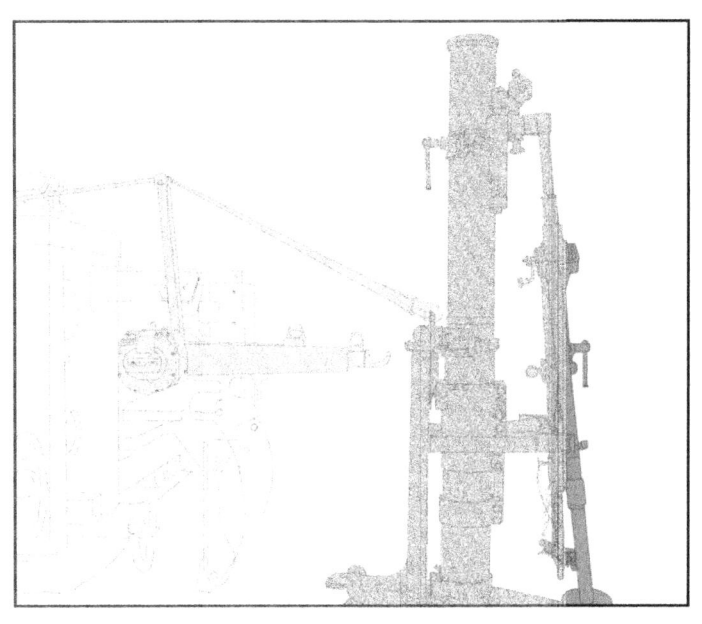

Figure 5-18. Mortar ready for lifting

- The ammunition bearer/driver moves the vehicle into position to allow the lift hook to engage the lift lug on the support strut. (See figure 5-19). Once the vehicle is in position, place it in PARK mode with the parking brake engaged. The ammunition bearer/driver mounts the trailer and awaits further instructions.

Figure 5-19. Lift hook in position

Note. The configuration of the support strut and hook allows for vehicle misalignment and maintains the ability to retrieve the mortar without the requirement of extremely accurate positioning of the vehicle.

- The gunner and assistant gunner lower the guide ramps.

Chapter 5

5-43. To skip to MANUAL mode, go to paragraph 5-45. In POWER mode:
- The assistant gunner manipulates the mortar assembly in the direction of the lift arm to allow engagement of the lift hook and to ensure the support strut engages lift hook.
- The gunner standing at the right rear of the M1101 trailer secures the control pendant. He pushes the on/off switch to the ON position. This engages the control buttons and powers the battery condition gauge. Check the voltage gauge on the front panel of the power enclosure to make sure there is enough power.

5-44. If there is no power, take the following troubleshooting steps:

Step 1. Check the fuze.

Step 2. Ensure the hydraulic pump unit switch is engaged.

Step 3. The assistant gunner places/ensures lift drive selector is in the POWERED position. The gunner lowers the lift arm to the best position to receive the mortar. The gunner commands, "ALL CLEAR," presses the UP button on the control pendant and continues to hold the button until the muzzle of the mortar tube is seated in the tube support assembly.

> **CAUTION**
> Ensure head and hands are clear of the barrel clamp assembly on the M326 120-mm mortar stowage kit when the M120-mm mortar is loaded.

Step 4. Once the barrel is lowered into the barrel clamp, the squad leader closes and locks the barrel locking clamp. Once the mortar is locked into the stowed position, the gunner pushes the DOWN button momentarily to relieve pressure on the lift arm, and turns off the control pendant.

Manual Mode

5-45. The assistant gunner ensures the lift drive selector (see figure 5-20) is in the MANUAL position and lowers the lift arm to the desired position. While facing the open end of the M1101 trailer, the ammunition bearer places his left hand on the winch drum and turns the handle in a counter clockwise direction. Once the handle is loose, the ammunition bearer maintains outward pressure on the handle. He grasps the winch strap and pulls it through the strap guide hole in the mortar support. (See figure 5-21).

120-mm Mortars

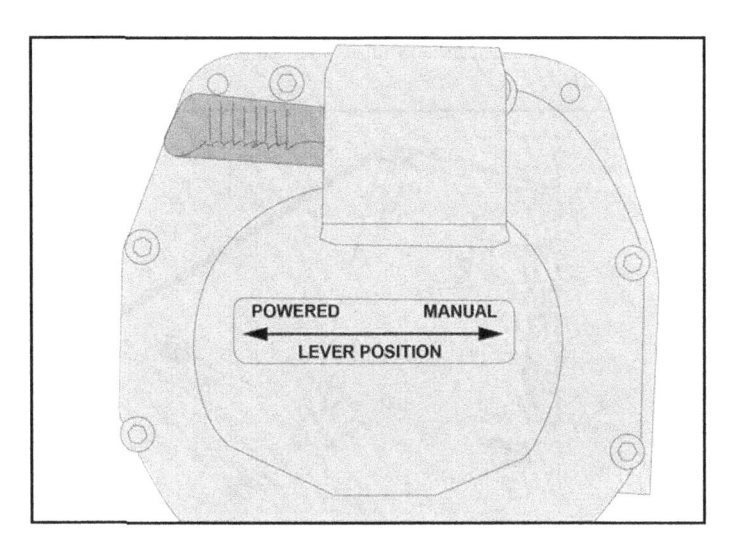

Figure 5-20. Lift drive selector

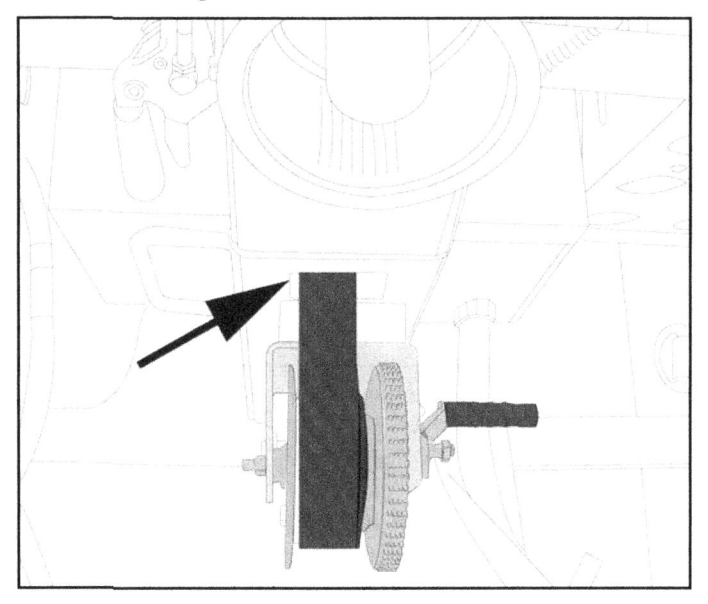

Figure 5-21. Winch strap guide hole

Chapter 5

5-46. The assistant gunner disconnects and raises the manual winch standoff to its upper most position. (See figure 5-22.)

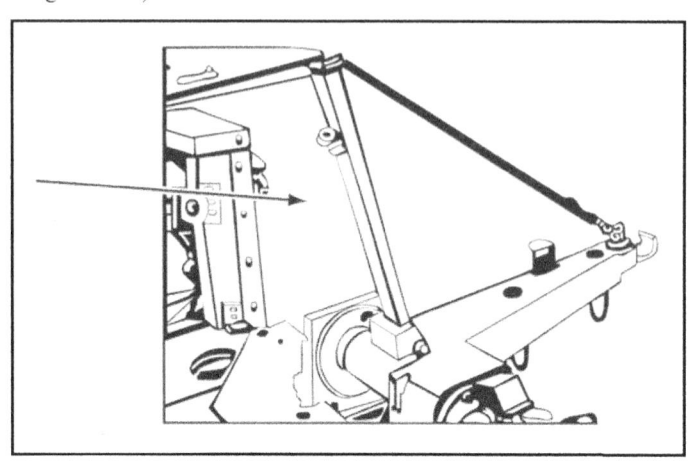

Figure 5-22. Manual winch standoff

5-47. Assistant gunner attaches winch strap to the manual lift arm. (See figure 5-21.) The assistant gunner pushes the mortar system towards the lift arm. The ammunition bearer turns manual winch handle in a clockwise direction, lifting the M120-mm system.

5-48. Once the barrel is lowered into the barrel clamp, the squad leader closes and locks the barrel locking clamp.

In POWER and MANUAL Modes

5-49. The squad leader disconnects 9w1 P1 cable, removes the FCC from the electronics rack, plugs the 9w1 P1 cable into the back of the vehicle, and installs FCC in the vehicle visor mount.

Note. If the trailer is at an upward angle rearward due to terrain, the baseplate may need to be pulled away from the trailer to help the strut rollers align with the guide ramps.

DISPLACING THE ELECTRONIC RACK

Note. The electronic rack is a three-man lift (gunner, squad leader, and assistant gunner).

5-50. The ammunition bearer/driver mounts the trailer and ensures the tie-down straps are not in the way. The gunner ensures the electronics rack is not connected to the NATO adapter at the junction box. The gunner secures one side of the upper electronics rack and the assistant gunner secures the other side of the upper electronics rack. In one fluid motion, they lift the electronics rack up to the side wall of the trailer with the assistance of the squad leader.

M1064A3 MORTAR CARRIER OPERATIONS

5-51. This section is a guide for training mortar units equipped with the M1064A3-series mortar carrier for mounting the M121 mortar. The procedures and techniques used for a mounted mortar are different from the ground-mounted mortar.

5-52. The M1064A3 carrier (see figure 5-23A and figure 5-23B) is an M113A3 armored personnel carrier modified to carry the 120-mm M121 mortar on a specially designed mount. It is fully tracked, highly mobile, and armor protected. It can be transported by air and is able to propel itself across water obstacles.

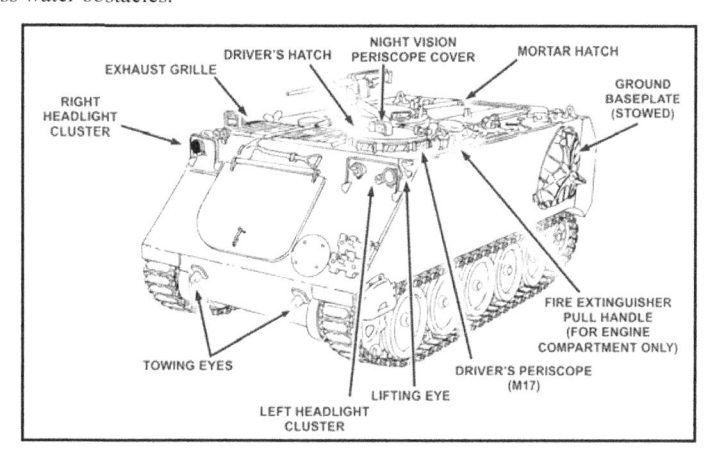

Figure 5-23A. M1064A3 mortar carrier, front and side

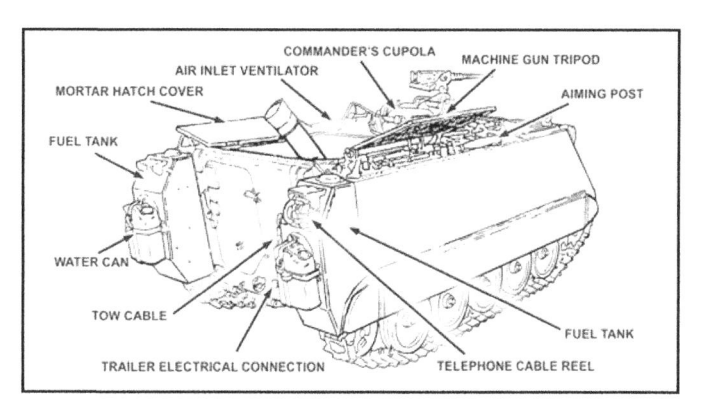

Figure 5-23B. M1064A3 mortar carrier, rear view

5-53. The mortar is mounted on its vehicular mount, and a clamping support is provided to hold the mortar during travel. A bipod support assembly provides attachment mounting for the M191 bipod and secures it in a locking position during travel. The breech socket provides a base in which the weapon rests. The mortar is provided with a sight extension arm assembly, which is received by the socket of the coupling and sight mount assembly. The gunner uses the sight extension arm to sight on the aiming point above the hull of the vehicle. The extension must be removed before moving the mortar to prevent wear on the sight mount's coupling gears.

Chapter 5

5-54. Care and cleaning of the mortar, instruments, and equipment are the duty and responsibility of the mortar squad. Care and cleaning of the carrier-mounted mortar are the same as for the ground-mounted mortar. All maintenance records and lubricating procedures for the 120-mm mortar and the M191 mount can be found in TM 9-1015-250-10. (Refer to TM 9-2350-277-10 for maintenance procedures for the M1064A3 carrier.)

OPERATION OF A CARRIER-MOUNTED 120-MM MORTAR

5-55. The mortar section is the fire unit for the mortar platoon. When a position is occupied, mortars are emplaced 75 meters apart, making the section front (distance between flank mortars) about 150 meters. The mortars are numbered 1, 2, and 3 (from right to left) when facing the direction of fire. The squad of the mortar carrier consists of four members—squad leader, gunner, assistant gunner, and ammunition bearer/driver.

Note. Digital systems, including the MFCS, allows for terrain positioning of the individual mortar guns within the platoon or section. The platoon sergeant/platoon leader or section leader should ensure that guns are separated, at a minimum, by the burst diameter of the round.

5-56. The premount checks for the carrier-mounted mortar are the same as in paragraphs 5-15 to 5-18 on pages 5-4 and 5-5.

PLACING A CARRIER-MOUNTED 120-MM MORTAR INTO ACTION

5-57. To place the mortar into action, the squad performs the following actions upon the squad leader's command, "ACTION:"

Note. If the weapon system cannot be leveled at any elevation given, the gunner is authorized to go to high range (low range) and move the buffer housing assembly along the cannon between the circular white line and the lower collar stop without moving past the vertical position, until the gun system can be leveled.

- The designated squad member pulls the chain on the center cargo hatch, releasing and folding it over onto the right hatch, ensuring that the hatch locks into place. He then pulls the chain on the right cargo hatch, releasing and folding it over, and secures it. Once the first squad member has the center and right cargo hatch secured, a second squad member pulls the chain on the left cargo hatch, releasing and folding the hatch over, ensuring that it is locked into place.

Note. The squad leader must look to ensure that the left and right sides of the vehicle are clear of obstructions (on top of the carrier, in path of the hatches) or personnel (on the right side of vehicle because of the double hatch).

- The assistant gunner unlocks the clamping support assembly.
- The gunner loosens the clamp handle assembly and grasps the buffer housing assembly, pulling it down until it is flush with the lower collar stop.
- The gunner ensures that the white line on the cannon aligns with the white line on the buffer housing assembly. He then tightens the clamp handle assembly until he hears a metallic click.

120-mm Mortars

- The assistant gunner then passes the sight extension to the gunner and removes the muzzle cover.
- The gunner places the sight extension into the dovetail slot on the bipod and secures it.
- The gunner mounts the sight onto the sight extension, places a deflection of 3200 mils and an elevation of 1100 mils onto the sight, and levels the deflection bubble with the cross-leveling handwheel on the bipod.
- The assistant gunner places the BAD onto the cannon, secures it, and levels the elevation bubble.
- The gunner then ensures that the traversing mechanism is within four turns of center, and the traversing extension is centered on the bipod.
- When the gunner is satisfied that the mortar is mounted correctly, he announces, "UP," so that the squad leader can inspect the weapon system.
- The carrier-mounted 120-mm mortar is placed into action and is ready for reciprocal lay.

Mounting the Mortar From a Carrier to a Ground-Mounted Position

5-58. The procedures for placing the mortar into action by mounting it from a carrier- to a ground-mounted position are described below:

Note. Left and right are in relation to the mortar's direction of fire.

- The driver lowers the ramp and dismounts from the mortar carrier.
- The gunner pulls down on the chain for the center hatch and folds it over, then secures it to the right cargo hatch. He pulls down on the chain for the right cargo hatch and folds both hatches over, then secures them in place. The assistant gunner pulls down on the chain for the left cargo hatch and folds it over, then secures it in place.
- The squad leader dismounts the carrier and shows the squad where he wants the mortar to be mounted, and indicates the direction of fire.
- The driver/ammunition bearer and assistant gunner release the baseplate by removing the safety pin and pushing the handle up. Together, they tilt the baseplate out and lift it from the lower brackets.
- The driver/ammunition bearer and assistant gunner place the baseplate at the firing position while the gunner retrieves the sight and aiming posts and places them on the left side of the baseplate.
- The assistant gunner holds the bipod while the gunner unlocks the clamp handle assembly. The assistant gunner lowers the bipod until it rests on the ramp. At the same time, the driver/ammunition bearer assembles the aiming posts.
- The assistant gunner releases the safety pins and rotates the handles until the arrows are facing each other, and then pulls the handles out, releasing the bipod.
- The assistant gunner installs the bipod leg extensions onto the bipod and secures them with the safety pins.
- The assistant gunner and driver/ammunition bearer secure the bipod and carry it to the firing position. They place it about two feet in front of the baseplate.

Chapter 5

> **CAUTION**
>
> Damage may occur to the turntable socket if the following procedures are not followed.

- The driver/ammunition bearer removes the muzzle plug while the gunner releases the clamping support assembly. Together, they raise the cannon to a 60-degree angle and rotate it until the white line is facing the turntable. They then lift straight up removing the cannon from the socket.

- The gunner and driver/ammunition bearer carry the cannon to the baseplate. With the white line on the cannon facing the ground, they tilt the cannon to a 60-degree angle and carefully insert the cannon into the baseplate socket. They rotate the cannon until the white line is facing skyward and then lower it onto the buffer housing assembly, ensuring the white line on the barrel aligns with the white line on the buffer housing assembly. The gunner then slides the buffer housing assembly down the barrel until it is flush with the lower collar stop on the cannon.

- The gunner tightens the clamp handle assembly until a metallic click is heard.

- The gunner places the sight on the weapon and indexes a deflection and elevation.

Performing Safety Checks on an M1064 Carrier-Mounted 120-mm Mortar

5-59. Specific safety checks must be performed before firing mortars. Most can be made visually. The gunner is responsible for physically performing the checks under the squad leader's supervision. Perform checks using the following guidelines and according to TM 9-1015-256-13&P, local policies and procedures, and unit SOP.

> *Note.* The buffer housing assembly may need to be adjusted to complete this task.

5-60. To perform safety checks on a carrier-mounted 120-mm mortar, the gunner ensures that—
- There is mask and overhead clearance.
- The breech assembly is locked in the turntable socket and the white line on the cannon bisects the white line on the buffer housing assembly.
- The M191 bipod assembly is locked to the turntable mount and the arrows on the mount handles are pointed down (vertical).
- The safety pins are installed.
- The buffer housing assembly is secured to the cannon by loosening the clamp handle assembly about a quarter of a turn, and tightening the clamp handle assembly until he hears a metallic click.
- The cross-leveling locking knob is hand-tightened.
- The bipod support assembly is locked in the high or low position and the safety pin is installed.
- The turntable is in the locked position.
- The cargo hatches are open and locked.
- The BAD locking knob is hand-tightened.

- The firing pin is properly installed.
- The bipod is not forward of the vertical position.

Performing Safety Checks on a Stryker Carrier-Mounted 120-mm Mortar

5-61. Specific safety checks must be performed before firing mortars. Most can be made visually. The gunner is responsible for physically performing the checks under the squad leader's supervision. Perform checks using the following guidelines and according to TM 9-2355-311-10-3-1, TM 9-2355-311-10-3-2, and TM 9-2355-311-10-3-3, local policies/procedures, and unit SOP.

5-62. To perform safety checks on a RMS6-L Stryker-mounted 120-mm mortar:
- Ensure all hatches are in the locked position with pins inserted.
- Check mask and overhead clearance.
- Check to ensure the BAD locking knob is hand tight.
- Check that bronze bushing on folding mechanism is in the locked position.
- Check that traverse quick release lever is engaged.
- Check to ensure fluid levels in the replenisher and recuperator are at the correct levels.
- Ensure M67 is properly seated in the dove tail slot.
- Ensure recoil path is clear of obstructions.
- Check that the firing pin is seated fully.
- Ensure that the mortar is in battery.
- Ensure retaining ring safety plunger is in the locked position.
- Ensure that antennas are tied down.

Mask and Overhead Clearance

5-63. To determine mask clearance, the gunner places an elevation of 0800 on the sight and lowers the cannon until the elevation bubble is level. He places his head against the breech cap and sights along the barrel to see if any obstructions are in front of the mortars. With the gunner's head still against the breech cap, the assistant gunner traverses the mortar through its full range of traverse (traversing extension) to ensure that no obstructions are in front of the mortar.

5-64. To determine overhead clearance, the gunner places an elevation of 1511 on the sight and raises the barrel until the elevation bubble is level. He places his head as close as possible to the breech cap (space is limited with the MFCS on both the M1064 and Stryker) and sights along the barrel for any obstructions in front of the mortar. With the gunner's head still against the breech cap, the assistant gunner moves the mortar through its full range of traverse (traversing extension) to ensure that no obstructions are in front of the mortar.

Note. If obstructions are found at any point in the full range of traverse or elevation, the mortar is not safe to fire. In a combat situation, however, it may be necessary to fire the mortar from that position. If this is the situation, traverse and elevate the mortar until it clears the obstruction and level the sight by using the elevation micrometer knob. Record the deflection and elevation where the mortar clears the obstruction and report this information to the FDC.

Chapter 5

PERFORMING SMALL DEFLECTION AND ELEVATION CHANGES ON A CARRIER-MOUNTED 120-MM MORTAR

5-65. The gunner receives deflection and elevation changes from the FDC in the form of a fire command. If a deflection change is required, it precedes the elevation change. To make the changes, he takes the following actions:

Note. Small deflection and elevation changes are greater than 20 mils but less than 60 mils for deflection and greater than 30 mils but less than 90 mils for elevation.

- The gunner sets the sight for deflection and elevation, then takes the following actions:
  - The gunner places the deflection on the sight by turning the deflection micrometer knob until the correct 100-mil deflection mark is indexed on the coarse deflection scale. He continues to turn the deflection micrometer knob until the remainder of the deflection is indexed on the deflection micrometer scale.
  - The gunner places the elevation on the sight by turning the elevation micrometer knob until the correct 100-mil elevation mark is indexed on the coarse elevation scale. He continues to turn the elevation micrometer knob until the remainder of the elevation is indexed on the micrometer scale.
- The gunner lays the mortar for deflection, then takes the following actions:
  - After the deflection and elevation are indexed on the sight, the gunner floats the elevation bubble.
  - The gunner turns the elevating hand crank to elevate or depress the mortar until the bubble in the elevation level vial starts to move. This initially rough lays the mortar for elevation.
  - The gunner looks through the sight and traverses to realign on the aiming post. He traverses half the distance to the aiming post, then cross-levels.
  - Once the vertical cross hair is near the aiming posts (about 20 mils), the gunner checks the elevation vial. If required, he lays for elevation again by elevating or depressing the elevating mechanism.
  - He makes final deflection adjustments using the traversing handwheel and cross-levels by traversing half the distance and cross-leveling.
  - When the vertical cross hair is within 2 mils of the aiming posts, all bubbles are leveled, and the sight is set on a given deflection, the mortar is laid.

Note. If the given deflection exceeds left or right traverse, the gunner may choose to use the traversing extension assembly. This gives an added number of mils in additional traverse to avoid moving the bipod. To use the traversing extension, the gunner pulls down on the traversing extension locking knob and shifts the cannon left or right. He relocks the traversing extension locking knob by ensuring it is securely seated.

PERFORMING LARGE DEFLECTION AND ELEVATION CHANGES ON A CARRIER-MOUNTED 120-MM MORTAR

5-66. The squad receives a fire command requiring a deflection change of more than 200 mils but less than 300 mils and an elevation change of more than 100 mils but less than 200 mils. The gunner must announce, "GUN UP," within 55 seconds. Time starts when the last digit of the elevation is given. The given deflection and elevation must be indexed, without error, and with the bubbles centered. The vertical crossline is within plus or minus two mils of an aligned or compensated sight picture. The traversing mechanism is within four turns of center traverse. The traversing extension is locked in the center position. The gunner receives deflection and elevation changes from the FDC in the form of a fire command. If a deflection change is required, it precedes the elevation change. To make the changes, the following actions are taken:

5-67. Upon receiving the fire command, the gunner places the data on the sight and elevates or depresses the mortar to float the elevation bubble. The following actions are taken in the specified situations:

- If in low range:
  - Gunner unlocks the low range support latch.
  - Gunner places his left hand on the bipod leg and his right hand on the traversing mechanism.
  - Assistant gunner places his right hand on the bipod leg and his left hand on the traversing mechanism.
  - Gunner and assistant gunner together raise the bipod into high range.
  - The assistant gunner inserts the high range-locking pin to secure the bipod.

5-68.
- If in high range:
  - Gunner and assistant gunner together lower the bipod into low range.
  - Gunner pulls the high range-locking pin.
  - Gunner places his left hand on the bipod leg and his right hand on the traversing mechanism.
  - Assistant gunner places his right hand on the bipod leg and his left hand on the traversing mechanism.
- The assistant gunner dead levels the elevation bubble.
- The gunner looks into the sight to determine if he has to shift the mortar, and announces to the assistant gunner to unlock the turntable.
- The gunner looks through the sight, and with the assistant gunner's help, shifts the mortar until the vertical cross hair is aligned with the aiming posts.
- The gunner cross-levels the deflection bubble and looks into the sight. If the vertical cross hair is still within 20 mils of the aiming posts, the gunner tells the assistant gunner to, "LOCK IT," (referring to locking the turntable).
- With the turntable in the locked position, the gunner continues to traverse and cross-level until the vertical cross hair is lined up with the aiming posts and the deflection bubble is level.
- The gunner levels the elevation bubble and rechecks the data, bubbles, and four turns of center.
- Gunner announces, "GUN UP."

TAKING THE MORTAR OUT OF ACTION (GROUND-MOUNTED TO M1064A3 CARRIER-MOUNTED

5-69. The procedures for mounting the mortar on the carrier from a ground-mounted position are described below:
- The squad leader commands, "OUT OF ACTION, PREPARE TO MARCH."

Chapter 5

- The gunner places a deflection and elevation from the TM for storage.
- The gunner removes the sight from the dovetail slot, places it in the sight case, and secures it on the carrier.
- The gunner then replaces the sight mount cover and snaps it closed.
- The ammunition bearer retrieves the aiming posts and the aiming post lights, and stows them on the carrier.
- The assistant gunner kneels down and secures the bipod, and lowers it to its lowest elevation leaving about one quarter of a turn of the shaft showing.
- The gunner opens the buffer housing assembly with his left hand while securing the cannon assembly with his right hand.
- The gunner and ammunition bearer rotate the cannon assembly until the white line is on the bottom. They raise the cannon to about a 60-degree angle and lift it from the baseplate socket. They carry the cannon assembly to the carrier.
- The gunner and ammunition bearer make sure the clamping support assembly is open. They ensure the bipod support assembly is in the low-range position, and the turntable is centered and locked.
- The gunner and ammunition bearer position the mortar cannon assembly just above the clamping support assembly with the white line facing down.
- The gunner and ammunition bearer carefully insert the breech cap ball into the socket. They raise the mortar cannon assembly and rotate it until the white line is in the up position. They install the muzzle plug and lower the mortar cannon assembly onto the clamping support assembly and lock it.
- The assistant gunner takes the bipod assembly to the carrier.
- The assistant gunner rests the bipod assembly on the carrier ramp with the cross-leveling mechanism pointing toward the gunner's position.
- The assistant gunner removes the pins and leg extensions from the bipod legs and stows them in the step.
- The assistant gunner removes the handles from the bipod support assembly and ensures the arrows are facing each other.
- The assistant gunner slides the bipod legs onto the attaching slot on the bipod support assembly.
- With the arrows facing each other and the holes matching, the assistant gunner inserts the handles through the bipod support assembly slot and legs. He turns the handles so the arrows point toward the bipod legs and attaches the safety pins.
- The assistant gunner slides the buffer housing assembly up along the cannon assembly until it stops. He tightens the clamp handle assembly until hearing a metallic click.
- The assistant gunner and ammunition bearer carry the baseplate to the carrier.
- The assistant gunner and ammunition bearer make sure the pin is removed and the handle is up.
- The assistant gunner and ammunition bearer lift the baseplate with the spades away from the carrier and place the two corners without handles inside the rims of the lower brackets.
- The assistant gunner and ammunition bearer tilt the baseplate toward the carrier and pull the handle down and lock it with the pin.
- The squad leader inspects to ensure all equipment is accounted for and properly secured.

RECIPROCAL LAY OF THE MORTAR CARRIER SECTION

5-70. To reciprocally lay the mortar, follow the procedures below:

- The aiming circle operator lays the vertical cross hair on the mortar sight and announces, "AIMING POINT THIS INSTRUMENT."

- The gunner refers the sight (using the deflection micrometer knob) to the aiming circle with the vertical cross hair splitting the lens of the aiming circle and announces, "NUMBER (number of gun) GUN, AIMING POINT IDENTIFIED."

- The aiming circle operator turns the azimuth micrometer knob of the aiming circle until the vertical cross hairs are laid on the center of the gun sight lens. He reads the deflection micrometer scales and announces (for example), "NUMBER (number of gun) GUN, DEFLECTION, TWO THREE ONE FIVE."

- The gunner indexes the deflection announced by the aiming circle operator on the sight and lays again on the center of the aiming circle lens by telling the driver to start the carrier. Looking through the sight, he tells the driver which direction to move the carrier (if necessary) and cross-levels. He ensures there is a correct sight picture (vertical cross hair splitting the aiming circle lens), and the elevation and deflection bubbles are leveled. Once this is accomplished, he announces, "NUMBER (number of gun) GUN, READY FOR RECHECK."

- The aiming circle operator gives the new deflection, and the process is repeated until the gunner announces, "NUMBER (number of gun) GUN, ZERO MILS (or one mil); MORTAR LAID."

- As the squad announces, "MORTAR LAID," the aiming circle operator commands the squad to place the referred deflection (normally 2800) on the sight and to place out aiming posts.

- The gunner only turns the deflection micrometer knob and indexes a deflection of 2800 mils on his sight without disturbing the lay of the mortar.

- The ammunition bearer runs with the aiming posts along the referred deflection about 100 meters from the mortar, dropping one post halfway (about 50 meters).

- Once the far aiming post is placed out as described, the gunner uses arm-and-hand signals to guide the ammunition bearer to place out the near aiming post.

- Once the correct sight picture is obtained, the gunner announces, "NUMBER (number of gun) GUN, UP."

M1129A1 STRYKER MORTAR CARRIER VEHICLE OPERATIONS

5-71. This section is a guide for training mortar units equipped with the M1129A1-series mortar carrier for mounting the RMS6-L mortar. The procedures and techniques used for a Stryker-mounted mortar are different from the other ground and carrier mounted mortar systems.

5-72. The SBCT Infantry battalion's primary mission is to close with the enemy by fire and maneuver to destroy, capture, or repel the enemy's assault by fire, close combat, and counterattack. These operations typically incorporate a primary base of fire provided by the respective platoon's weapon squads, and is supported, when possible, by the direct and indirect fires of supporting systems.

5-73. Regardless of the operational scenario, Stryker Infantry tactical operations are controlled and synchronized at the battalion level. They are executed by companies employing organic or attached combined arms elements, and are supported by organic and supporting fires/effects. While

Chapter 5

the SBCT Infantry battalion's main combat mission is to defeat or destroy enemy forces, it also provides U.S. Army combat operations with a predominant force for seizing, securing, retaining, and controlling terrain.

5-74. The 120-mm mortar is a traditional Infantry battalion asset. Attached to rifle companies as necessary by the battalion commander, the 120-mm mortar provides crucial fire support in combat operations. In addition to the four 120-mm mounted mortar carrier vehicles assigned to the battalion mortar platoon, each SBCT Infantry company is assigned a two-mortar carrier vehicle section consisting of 10 Soldiers that make up two mortar squads. The battalion mortar platoon as an organic asset of the Infantry battalion can perform direct support, general support, or reinforcing missions.

STRYKER MORTAR VEHICLE CHARACTERISTICS AND CAPABILITIES

5-75. The Stryker mortar vehicle's characteristics and capabilities are described in the paragraphs below.

5-76. The Stryker mortar vehicle (see figure 5-24) is a configuration of the Stryker Infantry carrier vehicle (ICV) variant. The Stryker mortar vehicle has the on/off-road ability of an ICV. While the power pack is the same, the Stryker mortar vehicle has a more robust suspension and is wider than the ICV. The rest of its characteristics are mortar-carrier specific. Stryker mortar vehicle squads provide mobile, high-angle fire for close-in support of ground troops in complex terrain and urban environments.

Figure 5-24. Stryker mortar carrier vehicle

SQUAD CONFIGURATION

5-77. The Stryker mortar vehicle carries a squad of five: driver, squad leader, ammunition bearer, gunner, and assistant gunner. (See figure 5-25.) Each squad member is assigned a duty station within the Stryker mortar vehicle.

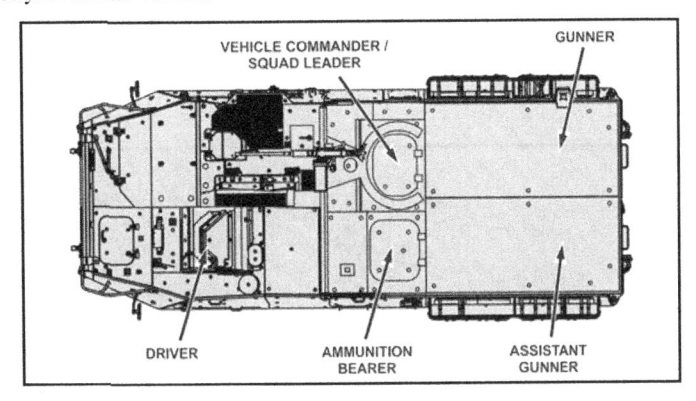

Figure 5-25. Squad configuration

5-78. An M95 mortar fire control system display was installed in the Stryker mortar vehicle's driver compartment to assist the **driver** with positioning the vehicle for firing the mounted mortar. An external step and handhold has also been added to aid entry and egress from the vehicle. There are no other differences in configuration or operation of the driver's station from the basic ICV. (Refer to TM 9-2355-311-10-3-1 for details on the Stryker mortar vehicle driver compartment.)

5-79. The Stryker mortar vehicle variant's absence of a remote weapons station (RWS) distinguishes its unique squad leader's station (see figure 5-24, page 5-42). The Stryker mortar vehicle doesn't have an RWS display, which allows for more space for the **squad leader** and a mounted video display terminal. The video display terminal can display the driver's thermal viewer and Force XXI Battle Command, Brigade-and-Below (FBCB2) screens. The commander's interface (CI) display and FBCB2 display are mounted on swing-out mounts to enable viewing from both squad leader and ammunition bearer positions.

5-80. The **ammunition bearer** sits to the left of and facing toward the squad leader. (See figure 5-24.) His station includes a single seat with a lap belt restraint, a padded base, and a fold-down backrest. The CI display and FBCB2 display swing out for use from this position.

Note. The M3 heater and ventilated facemask hose have been moved from the ICV common variant position. Both are mounted forward and left of the ammunition bearer's position on the transition of the engine bulkhead.

5-81. The **gunner/assistant gunner** seats are in the rear of the troop compartment. (See figure 5-24.) The gunner sits on the right-rear side facing toward the front of the vehicle; the assistant gunner sits on the left-rear side of the vehicle facing the cannon's blast attenuator device. Both station's seats are designed to provide passenger restraints, including lap belts with inertia-type retractors. Both stations are equipped with overhead handholds (subway straps) mounted to the roof of the vehicle. In addition, the gunner and assistant gunner have their own M3 heater and ventilated facemask hose that is located above the radio rack on the right side sponson shelf.

Chapter 5

Weapons Systems

5-82. The Stryker mortar vehicles are equipped with two weapon systems: an M240B machine gun mounted at the commander's station and a mounted RMS6-L 120-mm mortar. (See figure 5-26.) The Stryker mortar vehicle is designed to carry an 81-mm mortar (battalion load plan), and a 60-mm mortar (company load plan), with ammunition storage capabilities for all three systems.

5-83. The Stryker mortar vehicle retains the same integral protection characteristics as the ICV. However, it is not provided with slat armor at initial delivery, though it can be mounted with slat armor, if required.

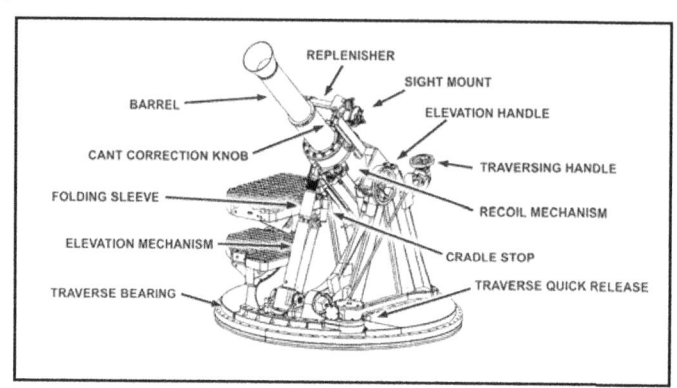

Figure 5-26. RMS6-L 120-mm mortar

Carrier Doors

5-84. The two mortar doors on the roof of the Stryker mortar vehicle span the length of the troop compartment; from just behind the commander and ammunition bearer—to the rear of the vehicle. (See figure 5-27.) Each door weighs about 400 pounds with an opening/closing force of about 65 pounds.

5-85. The latches to open/close the doors remain ICV-common to other latches on the vehicle. However, the latch mechanism to hold the doors in the open position for firing is unique to the Stryker mortar vehicle. The latch design securely holds the doors in the open position while the vehicle is being repositioned. A Stryker mortar vehicle unique locking pin secures the doors in the open position.

Figure 5-27. Mortar doors

CAUTION

Moving with the mortar doors open and latched is limited to short distances and vehicle speed no faster than 10 mph. Squad members must verify that the mortar doors are latched and the latch pins are properly installed. Squad members must remain in their seats with their seatbelts properly secured. The 120-mm mortar must be secured in the travel position.

WARNING

Mortar doors are large and extremely heavy. Two Soldiers are required to safely open, close, and secure the doors in the opened or closed positions. Attempts to open or close mortar doors with less than two people may result in serious injury.

Mortar doors must be securely locked when in the open position, with safety pins installed to prevent doors from accidentally closing. Failure to lock mortar doors in the open door position may result in serious injury.

WARNING

Mortar doors must be securely locked when in the closed-door position. Doors may bounce up and down while vehicle is moving and cause serious injury.

WARNING

Mortar doors are under spring tension. To avoid serious injury, keep face and body parts clear of doors, and hands clear of door edges when opening or closing mortar doors.

5-86. A minor modification made to the left rear of the Stryker mortar vehicle accommodates a horizontal ammunition rack. (See figure 5-28.) This adaptation incorporated changes to the upper sidewalls of the Stryker mortar vehicle that restructured the common upper sidewall angles of the ICV to a vertical angle. To accommodate the ammunition rack, portions of the spall lining were removed. More armor was added to this area to retain the same level of protection.

Figure 5-28. Horizontal ammunition rack

Stryker Mortar Vehicle Configurations

5-87. The Stryker mortar vehicle recoiling mortar system incorporates some characteristics of the basic M120/M121. Except for adding an external shoulder on the breech cap that allows mounting the barrel into the recoiling system, the mortar tube is identical to the M120/M121. The RMS6-L 120-mm mortar system recoiling mechanism is designed to reduce the recoil forces transferred to the vehicle. The type of dismountable mortar carried on the vehicle is dependent upon vehicle configuration. Configurations include:

- **Battalion configuration.** The platoon Stryker mortar vehicles consists of the 120-mm mounted mortar system and carries the 81-mm mortar M252/M252A1 for dismounted use. The 81-mm base plate is stowed on the outside right rear of the vehicle above the tires. The bipod is stowed in the right side of the vehicle just behind the squad leader's station. The bipod can also be stowed on top of the vehicle just behind the winch pocket.
- **Company configuration.** The company section Stryker mortar vehicle consists of the 120-mm mounted mortar system and carries the 60-mm mortar M224/M224A1 for dismounted use. The 60-mm base plate, bipod, and cannon are stowed in the same location as the 81-mm on the battalion configuration.

Reconnaissance, Surveillance, Targeting and Acquisition Squadron

5-88. The SBCT battalion's reconnaissance, surveillance, targeting, and acquisition (RSTA) squadron Stryker mortar vehicle consists of only the 120-mm mounted mortar system.

Note. The Stryker mortar vehicle configurations are driven by the different assignments of the Stryker mortar vehicle to the battalion, company, or RSTA squadron.

5-89. The secondary weapon on the Stryker mortar vehicle is the M240B 7.62-mm machine gun. The machine gun is attached to a skate mount with an articulated arm. A total of 1000 rounds of

Chapter 5

ammunition can be stowed on the exterior of the vehicle. The stowage box is to the right of the commander and can hold up to five cans (200 rounds each) of ammunition.

RMS6-L 120-MM MORTAR SYSTEM

5-90. The RMS6-L 120-mm mortar system is a simply constructed and mobile weapon system that can be brought into or taken out of action quickly. It is capable of producing a large volume of fire quickly and accurately on any given target within its range. The system is designed for employment in all phases and types of land warfare, on every type of terrain, and in all kinds of weather.

Stryker Mortar Vehicle 120-mm Mortar Capability

5-91. The mortar for the Stryker mortar vehicle is a smooth bore, muzzle loaded, high angle of fire weapon that provides—
- A traverse field of 4400 mils.
- A maximum range of 6700 meters.
- A maximum angle of elevation of 1486 mils, enabling the mortar to engage targets effectively on reverse slopes or behind cover.
- Accurate firing on targets from 180 to 6700 meters.

Stryker Vehicle 120-mm Mortar Characteristics

5-92. The mortar for the Stryker mortar vehicle system includes—
- A smooth bore barrel with rounded muzzle end to allow for easy loading.
- A sight instrument mounting bracket with cant correction knob provided for mounting the M67 sight unit.
- A replenisher mounted on top of the cradle that contains hydraulic fluid for the recoil system. The cradle assembly is attached to the saddle and contains a recoil mechanism that buffers barrel recoil during firing.
- Trunnions that allow for vertical pivot of the cradle.
- A connecting hub that attaches the barrel assembly to the cradle.
- The breech piece, which is screwed into the end of the barrel to form a gas-tight metal seal. This breech piece houses the firing pin and safety mechanism.
- A traversing handle that provides quick and accurate rotation of the turntable changing the barrel azimuth.
- A saddle assembly, which supports the cradle/barrel assemblies and attaches to the traverse bearing of the lower vehicle mount.
- A traverse bearing assembly that attaches the mortar and recoil system directly to the Stryker mortar vehicle.
- A quick release lever used to disengage the traversing gear from the traverse bearing assembly, allowing free rotation of the saddle assembly.
- An elevating mechanism attached through a folding mechanism to the underside of the cradle, which allows lowering the cannon to the travel position.
- An elevating handle that provides quick and accurate raising or lowering of the cannon in elevation.
- A folding mechanism bushing that allows lowering of the cannon to the travel position.
- A recuperator for maintaining nitrogen and fluid pressure forces the gun back into battery after recoil.

120-mm Mortars

Note. To accept the weight and recoil forces transferred to the vehicle by the RMS6-L Stryker mortar vehicle 120-mm mortar system, slight modifications were made to the floor panels. A mounting pad was added to provide for additional support and mounting of the bearing (turntable) on which the mortar traverses. The main bearing of the mortar system is bolted directly to the mounting pad on the floor of the Stryker mortar vehicle. A fuel transfer control access cover was added to the floor plates.

Mortar Ammunition Storage

5-93. The Stryker mortar vehicle's permanent rack on the left side of the Stryker mortar vehicle holds only 120-mm rounds in horizontal- and vertical-ready round racks. (Refer to chapter 1 for detailed instructions on mortar ammunition stowage and removal.) The ammunition storage system also consists of a modular rack system on the right side of the vehicle capable of storing 60-mm, 81-mm, and 120-mm ammunition. Storage is possible in three different combinations depending on the vehicle configuration (battalion, company, or RSTA).

5-94. The Stryker mortar vehicle is capable of storing 60 120-mm rounds stored within their individual round containers. The left rear side of the vehicle contains a 120-mm rack that is capable of storing 48 120-mm rounds: 24 horizontal and 24 vertical. The right side has a flex rack that is capable of storing 12 rounds of 120-mm mortar ammunition. The right side flex rack also is capable of storing 35 81-mm rounds. On the 120-mm and 81-mm flex rack located in the right side of the Stryker mortar vehicle, mortar rounds are held in place by using webbed straps that tighten around the ammunition to provide a positive means of holding the rounds in place. (See figure 5-29, page 5-50.)

5-95. The Infantry company Stryker mortar vehicle is capable of storing both 120-mm and 60-mm rounds. The left side storage space is common to the battalion and RSTA squadron's Stryker mortar vehicles. The right-side rack is capable of storing 77 rounds of 60-mm ammunition. The 60-mm flex rack incorporates a single door that covers the front of the rack. In addition to the webbed straps and aluminum door of the 60-mm configuration, there are wave springs that keep rounds from sliding or rolling while loading and unloading.

5-96. The RSTA squadron Stryker mortar vehicle carries the same number of 120-mm rounds as the battalion and company configurations. Though RSTA Stryker mortar vehicles carry only 120-mm mortar rounds, they can be configured to stow 60-mm and 81-mm ammunition.

Figure 5-29. Right side ammunition rack

MANUAL BORESIGHT PROCEDURES USING THE M45-SERIES BORESIGHT FOR THE RMS6-L 120-MM MORTAR

5-97. Boresighting is a vital accuracy-enhancing measure that must be conducted during every long halt, in assembly area procedures, and in consolidation and reorganization.

5-98. In preparation for the manual calibration of the M45 boresight of the Stryker mortar vehicle's equipped 120-mm mortar, the vehicle should be positioned on flat level terrain.

5-99. The mortar must be oriented onto an aiming point at least 200 meters away. It must have a clearly defined vertical line (see figure 5-30), and be placed into action with the mortar barrel assembly in the center of traverse.

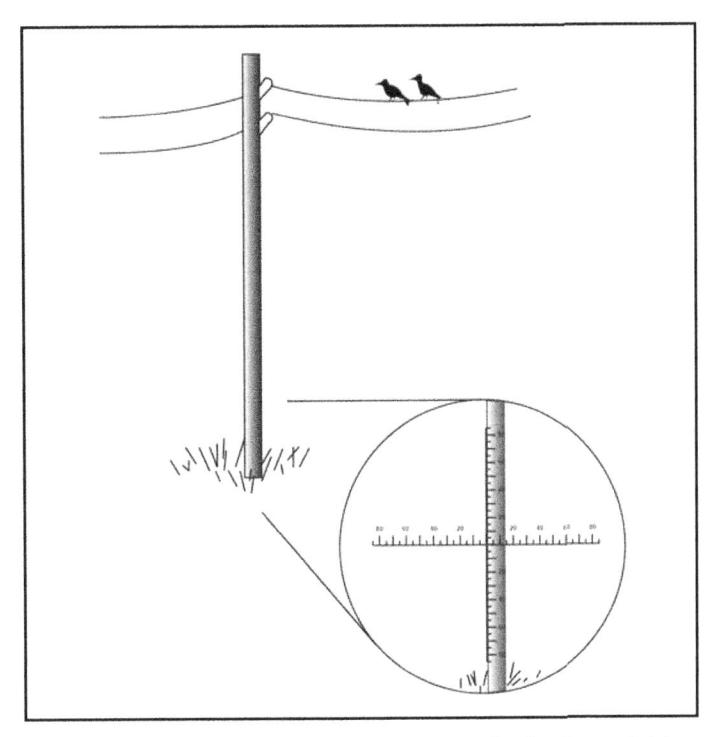

Figure 5-30. Example distant aiming point for boresight

MANUAL BORESIGHT SETUP

5-100. The following set-up procedures are required to manually boresight the 120-mm mortar:

Step 1. Install the M67 sight unit (see figure 5-31 on page 5-52) and index a deflection of 3200 mils and an elevation of 800 mils.

Step 2. Place the elbow telescope eyepiece parallel to the ground and align index marks on the telescope and eyepiece. Lock the eyepiece in place using the wingnut on the loop clamp.

Chapter 5

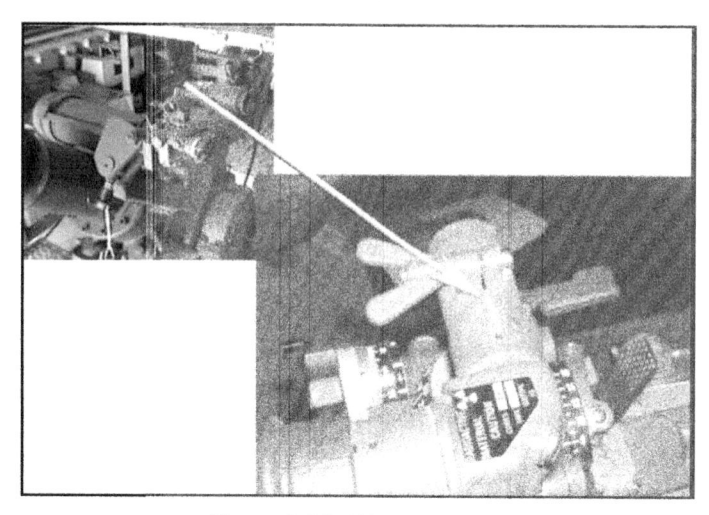

Figure 5-31. M67 sight unit

Step 3. Remove cant, if present, by turning the cant correction (cross-level) knob (see figure 5-32) until the bubble in the cross-level on M67 sight is centered.

Figure 5-32. Cant correction (cross-level knob)

Step 4. Release the telescope lock and move the telescope vertically as necessary, until the aiming point is visible.

Step 5. Look through the elbow telescope and direct the driver to maneuver the vehicle until the vertical crosshair is within 20 mils of the distant aiming point. Do not use the traversing mechanism. Ensure the deflection and elevation are set correctly.

Step 6. Remove the BAD (see figure 5-33). Inspect the cannon to ensure mounting surface is free of any burrs or projecting paint imperfections that would prevent the boresight device from being mounted properly.

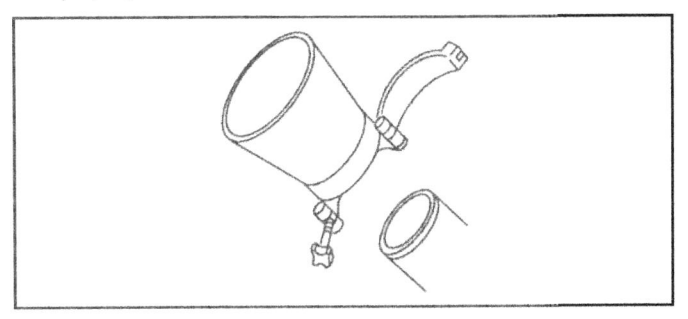

Figure 5-33. Blast attenuator device

Set the Elevation for Boresight

5-101. The following set-up procedures are required to set the elevation for the boresight:

Step 1. Remove the M45A1 boresight device and mounting straps from the carrying case, and place the boresight on top of the mortar barrel assembly just below and against the upper stop band with the elbow telescope in the direction of the gunner. (See figure 5-34.)

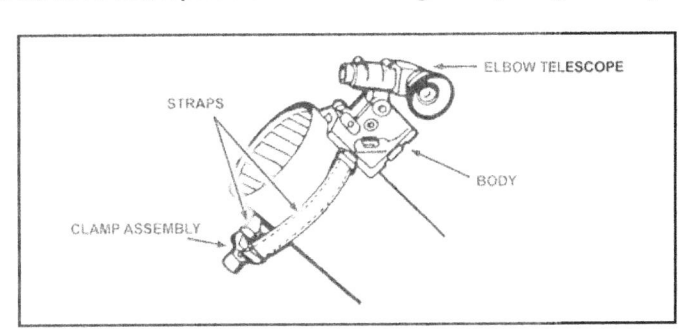

Figure 5-34. M45 boresight on cannon

Step 2. Center the cross-level bubble by rotating the boresight slightly on the outside diameter of the cannon.

Note. Loosening the clamp screw and lightly tapping the boresight body allows slight movements around the cannon. When the cross-level bubble centers, tighten the clamp screw.

Step 3. Elevate or depress the cannon until the boresight elevation bubble is centered (see figure 5-35) if necessary, cross-level the boresight. The mortar now is set at 0800 mils elevation. The vertical hairline of the boresight should be within 20 mils of the aiming point. If not, repeat steps 1 through 3 using the boresight. Complete all steps above, so the vertical hairline of the boresight is within 20 mils of the aiming point.

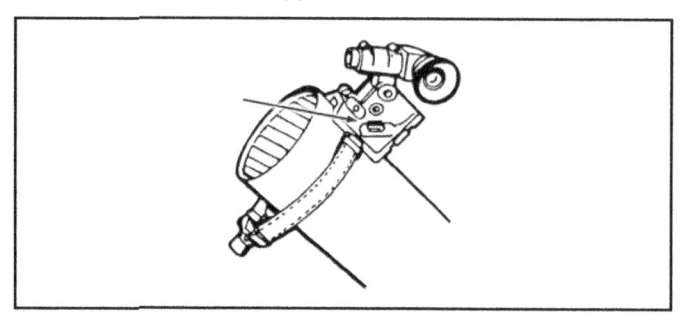

Figure 5-35. M45 boresight elevation bubble

Note. Always sight along the left edge of the aiming point.

Step 4. Check cross-level vial on the M67 sight unit. Center the bubble by turning the tilt correction (cross-level) knob until the bubble is centered.

Step 5. If necessary, repeat steps 1 through 4 until the bubbles in the cross level and elevation vials of the M67 sight, and cross-level and elevation vials of boresight, are centered.

5-102. The reading on the course elevation scale of the M67 sight should be 800 mils and the reading on the fine elevation micrometer scale should be zero. If adjustment is necessary, proceed with the following steps:

Step 1. Lock the fine elevation micrometer knob using the red elevation micrometer locking knob. Loosen the two elevation micrometer knob set screws and slip the elevation micrometer scale until the 0 mark on the micrometer scale coincides with the index arrow on the housing. Tighten the two screws to secure the micrometer scale.

Step 2. Once the micrometer knob has been properly indexed at zero, check the reading on the coarse elevation scale. If 800 mils is not precisely in line with the index arrow, unlock and rotate the micrometer knob to achieve proper alignment. The micrometer knob scale should read plus or minus 20 mils of the zero. If the M67 sight unit exceeds the 20 mil tolerance allowed, the sight must be turned in for calibration. Movement of the coarse elevation scale is not authorized.

Step 3. Ensure all bubbles are level and adjust, as necessary.

Step 4. The mortar now is sighted for elevation.

Set the Deflection for the Boresight

5-103. The following set-up procedures are required to set the deflection for the boresight:

Step 1. Look in the boresight and determine if the mortar must be shifted to the target again. If the mortar needs to be shifted, traverse half way to the target, ensuring you are within four turns of center traverse, and tap level the deflection bubble on the boresight (repeat this step until the boresight vertical crossline is lined up on the left edge of the target).

Step 2. Adjust the boresight and the M67 sight unit to keep the cross-level bubble since the mortar may cant as it is traversed. It may be necessary to center the elevation bubble on both instruments.

Step 3. Rotate the deflection micrometer knob until the M67 sight unit's vertical hair line is aligned on the left edge of the aiming point.

Step 4. The coarse deflection scale should read 3200 mils. The deflection micrometer scale should read 0. If adjustment is necessary, proceed with steps A through C below:

Step A. Lock the deflection micrometer knob using the red deflection micrometer locking knob. Loosen the two deflection micrometer knob set screws and slip the red line on the deflection micrometer control dial to align with the red arrow on the dial pointer. Recheck and if necessary, adjust the sight picture. Then tighten the two screws to secure the control dial. Once again, push the black deflection micrometer scale toward the body of the M67 sight unit and rotate the micrometer scale to align the 0 with the black arrow on the dial pointer.

Step B. Gently push down on the deflection coarse scale and rotate to align the 3200 mil graduation with the index arrow on the sight unit body. If the red line on the dial pointer and the 3200 mil graduation are in alignment, the deflection setting is complete. If the red line on the dial pointer and the 3200 mils graduation is more than 20 mils out of alignment, it must be adjusted.

Note. To check the alignment, align the 3200 mil graduation with the red line on the dial pointer. Measure the mil difference on the deflection micrometer knob by unlocking and rotating the deflection micrometer knob until the 3200 graduation is in alignment with the index arrow and reading the mil difference from 0.

Step C. If necessary, with the deflection micrometer knob and coarse scale correctly indexed to 3200 mils and locked, loosen the dial pointer set screws and rotate the red line on the dial pointer until it is in alignment with the 3200 mil graduation. Tighten the two screws to secure the dial pointer. Unlock the deflection micrometer knob.

Step 5. Recheck and level all bubbles. Ensure all data is precisely set, control dial and dial pointer are within tolerances, both sight pictures are identical on the aiming point, and the turntable is no more than 20 mils from center. If any or all of these conditions are not met, repeat the necessary boresight procedures and recheck.

Correcting for Cant

5-104. The following set-up procedures are required to correct for cant:

Step 1. To ensure proper alignment, remove and place the boresight in position underneath the cannon. Center the boresight cross-level bubble and check the vertical crosshair to see if it is still on the aiming point. If cant exists, the vertical crosshair of the boresight will not be on the aiming point. This indicates that the true axis of the bore lies halfway between the aiming point and where the boresight is pointing.

Step 2. To correct this error, look through the boresight. Traverse the mortar onto the aiming point while keeping all bubbles level on both the boresight and sight unit. Using the deflection micrometer knob, place the vertical crosshair of the sight back onto the aiming point. Determine the mil difference from the original 3200 mils deflection and rotate the micrometer one half the mil difference back in the direction of 3200 mils. Lock the red deflection micrometer knob and continue to follow the steps outlined above in step 4 (A) of paragraph 5-103 to adjust the deflection micrometer knob back to 0.

Chapter 5

Step 3. Ensure all bubbles are level and the boresight retains the correct sight picture.

Step 4. Remove and store the M45 boresight and M67 sight unit in their carrying cases.

Step 5. Install BAD.

Step 6. Manual boresight is complete.

DIGITAL BORESIGHT PROCEDURES USING THE M45-SERIES BORESIGHT FOR THE RMS6-L 120-MM MORTAR

5-105. For the MFCS software to properly compute the boresight values, the mortar trunnions must be leveled. In a tactical environment, this is most easily accomplished using a gunner's quadrant.

5-106. To make the process easier to perform, the vehicle should be parked on a slight incline, roughly 20 to 50 mils, with the front of the Stryker mortar vehicle higher than the rear. The incline may be measured with the gunner's quadrant placed on top of the mortar's traversing mechanism gearbox, or traversing gear ring, and aligned with the fore and aft axis of the carrier. The LINE OF FIRE indicated on the gunner's quadrant (see chapter 2, figure 2-9 on page 2-17) should point toward the front of the vehicle.

Note. When using the gunner's quadrant, be sure the mating surfaces are free of any burrs, projections, or paint that would prevent the quadrant from seating properly. The gunner's quadrant must be held against the three registration pins on the seating pads to achieve an accurate reading.

Manual Boresight Setup

5-107. The following procedures are required for manual boresight setup:

Step 1. Ensure the MFCS is fully initialized and the navigation subsystem is fully aligned.

Step 2. Record the current easting, northing, and altitude position data. This information may be required at the end of the boresight procedure.

Step 3. Elevate the weapon to roughly 0800 mils as indicated by the gunner's quadrant placed on the mortar's elevation seats. (As the turntable is not level, the elevation will change as the weapon is traversed.)

Step 4. Set the gunner's quadrant to 0 mils plus or minus any correction value determined during the end for end test. The correction value should be recorded on the outside of the quadrant case.

Step 5. Place the gunner's quadrant on the mortar's 0800 mil cross-level seats and traverse the weapon to center the bubble. When moving the cannon, always make the last turn of the traversing handwheel in the direction of the most resistance.

Step 6. When the bubble is centered, remove and reseat the gunner's quadrant on the cross-level seats and verify the bubble is still centered. If not, repeat the prior step. If the bubble cannot be centered, relocate the vehicle to a more suitable position and repeat the previous step.

Step 7. On the commander's interface, select the pointing device control button and then select the boresight tab.

120-mm Mortars

Boresight at 0800 mils

5-108. For a boresight at 0800 mils, the following procedures are required:

Step 1. Set the gunner's quadrant to 0800.0 mils, plus or minus any correction value it may have.

Step 2. Place the gunner's quadrant on the mortar's elevation seats. Elevate or depress the cannon to center the bubble. When elevating the cannon, always make sure the last turn of the handwheel moves the cannon up. If 0800.0 mils is an overshot, lower the elevation to below 0800.0 mils and increase upward movement to center the bubble.

Step 3. When the bubble is centered, remove and reseat the gunner's quadrant on the seats at least two times and verify the bubble is centered both times. Consistency of this measurement is critical for boresight. Do not proceed to the next step until an accurate and consistent 0800.0 mil elevation (plus or minus any correction value) is achieved.

Step 4. Set the gunner's quadrant to zero mils.

Step 5. Place the gunner's quadrant on the mortar's 800 mil cross-level seats and verify the bubble is still centered. If not centered, traverse the weapon in the same direction needed to center the bubble.

Step 6. When the bubble is centered, remove and reseat the gunner's quadrant on the 800 mil cross-level seats and verify the bubble is still centered.

Step 7. On the commanders' interface, select the ON DAP AT 800 MILS button on the boresight tab.

Step 8. Record the azimuth displayed in the upper right-hand corner of the CI.

Boresight at 1300 mils

5-109. For at boresight at 1300 mils, the following procedures are required:

Step 1. Using the current elevation value, in the upper right hand corner of the CI, elevate the cannon slightly less than 1300 mils (approximately 1290 mils).

Step 2. Set the gunner's quadrant to 1300.0 mils plus or minus any correction value it may have.

Step 3. Place the gunner's quadrant on the mortar's elevation seats (coarse scale closest to the elevation seats). Elevate the cannon to center the bubble. When elevating the cannon, always make sure the last turn of the handwheel moves the cannon up.

Step 4. When the bubble is centered, remove and reseat the gunner's quadrant on the seats at least two times and verify the bubble is centered both times.

Step 5. Set the gunner's quadrant to zero mils.

Step 6. Place the gunner's quadrant on the mortar's 1300 mil cross-level seats and verify the bubble still is centered. If the bubble is not centered, traverse the weapon in the same direction needed to center the bubble.

Step 7. When the bubble is centered, remove and reseat the gunner's quadrant on the cross-level seats and verify the bubble is still centered. If not, repeat steps 1 through 7. Do not proceed to the next step until an accurate and consistent zero mil cross-level measurement is achieved.

Step 8. Compare the azimuth currently displayed in the upper right-hand corner of the CI to the previously recorded azimuth. The difference between the two should not exceed three mils.

Step 9. On the CI, select the ON DAP AT 1300 MILS button on the pointing device boresight tab.

Step 10. Read and record the azimuth, elevation, and roll values displayed in the NEW BORESIGHT CORRECTIONS window at the bottom of the pointing device boresight tab.

Step 11. Select the SAVE CORRECTIONS button on the pointing device boresight tab. PD OUT should be displayed in the upper right hand corner of the CI.

Chapter 5

Step 12. Set the PD POWER switch on the personal digital assistant to OFF. Ensure the precision lightweight GPS receiver (PLGR) or defense advanced GPS receiver (DAGR) is operational and after 10 seconds, switch the PD POWER to ON.

Note. If the PLGR or DAGR is not tracking the required number of satellites, the pointing device will not begin aligning and a manual position update is required, or you may wait to obtain the correct number of satellites so the pointing device aligns. Allow no movement on the vehicle during alignment.

Step 13. Select the PD STATUS tab and verify that the pointing device has restarted (by watching the countdown at the top right of the screen). If it has not restarted, select POSITION UPDATE and enter the current easting, northing, and altitude position data recorded during boresight set-up and press SEND UPDATE.

Step 14. When the alignment countdown ends, select the boresight tab.

Step 15. Verify that the azimuth, elevation, and roll values displayed in the current boresight correction window at the top of the pointing device boresight tab match the recorded values. If not, the values were not retained in the pointing device and the process must be repeated.

Step 16. Digital boresight now is complete.

Chapter 6

Mortar Fire Control System

The MFCS provides a complete, fully-integrated, digital, onboard fire control system for the carrier-mounted 120-mm mortar. It provides a "shoot and scoot" capability to the carrier-mounted mortar. With the MFCS, the mortar FDC continues to compute fire commands to execute fire missions and controls its gun carriers. The carrier-mounted MFCS components work together to compute targeting solutions, direct the movement of vehicles into firing positions, allow real-time orientation, and present gun orders to the gunner.

MFCS CAPABILITIES

6-1. The MFCS is an automated fire control system designed to improve the command and control of mortar fires and the speed of employment, accuracy, and survivability of mortars. The FCC, controlled by a software operating system, manages computer activities, performs computations, and controls the interface with peripheral and external devices. The FCC operator enters data at the keypad and composes messages using the liquid crystal display (LCD). Completed messages then are transmitted digitally or by radio.

6-2. Should the FDC become disabled, each mortar squad can compute its own fire missions if the FDC is configured as a gun/FDC. System accuracy is increased through the use of a GPS, an onboard azimuth reference for the gun, and digital meteorological data updates. The MFCS (see figure 6-1, page 6-3) provides the capability to have self-surveying mortars, digital CFF exchange, and automated ballistic solutions.

6-3. The 120-mm mortar system is aligned on the MFCS to maintain alignment with accuracy of 3 mils azimuth and one-mil elevation in all conditions. The commander's interface stores up to—
- Eighteen gun positions.
- Three gun sections.
- Fifty known targets.
- Sixteen registration points.
- Three final protective fire missions.
- Twelve forward observer locations.

6-4. If the designated fire direction center (FDC) becomes inoperative for any reason during tactical or training operations, any MFCS-equipped M1064 or Stryker mortar vehicle can assume FDC responsibilities.

MORTAR FIRE CONTROL SYSTEM MOUNTED

6-5. The M95 MFCS provides fire and maneuver capability to the carrier mounted mortar. Controlled by a software operating system, the commander's interface microprocessor manages

Chapter 6

computer activities, performs computations, and controls the interface with other components and devices. MFCS components work together to—
- Compute targeting solutions.
- Direct movement of the mortar vehicle into position.
- Allow real-time gun orientation.
- Present gun orders to the M95 components mounted in the Stryker mortar vehicle.

M95 MFCS COMPONENTS

6-6. MFCS components make up a complete, fully integrated, digital on-board fire control system that can establish weapon location and orientation without the use of a sight unit, aiming posts, or distant aiming points. The MFCS delivers seamless integration of mortar fires into the digital fire support network, calculates navigation instructions, and computes ballistic solutions while the carrier is moving.

6-7. Major components of the MFCS include the commander's interface, power distribution assembly, gunner's display, pointing device, and driver's display. They are each described as follows:
- Commander's interface capabilities include:
  - Managing the information flow between the gun and FDC.
  - Providing interface between MFCS components using text and graphics.
  - Computing technical fire control solution for weapon operation.
- Power distribution assembly filters Stryker mortar vehicle power through the direct current (DC) power system that isolates MFCS components from power fluctuations, and provides circuit breakers for MFCS components.
- Gunner's display receives deflection and elevation orders, and provides check fire and call for fire! capability to the gunner.
- Pointing device provides effective pointing and position performance between 80 degrees south to 84 degrees north latitude.
- Driver's display provides steering directions, distance and heading in a numerical format, and information needed to orient the Stryker mortar vehicle upon emplacement.
- Vehicle motor sensor.

Note. Refer to TM 9-1220-248-10 and FM 3-22.91 for more information about the MFCS.

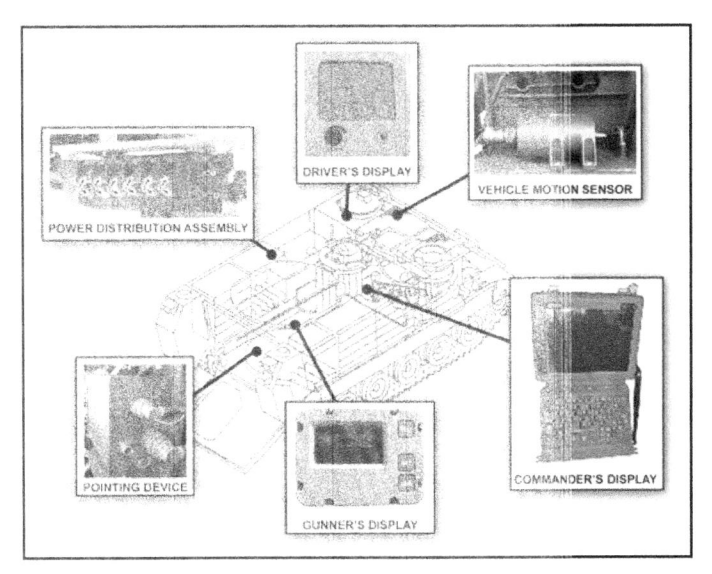

Figure 6-1. Mortar Fire Control System

6-8. The **commander's interface** (see FCC, figure 6-2, page 6-4) is a computer terminal that provides high-speed data processing, an LCD, and a QWERTY-style keyboard similar to a common personal computer. It has a built-in modem allowing a wide variety of data exchange requirements. The operator works with the FCC graphic user interface (GUI) to operate the system by using the built-in mouse or keyboard, and pointing and clicking buttons and tabs. Data is presented on screens designed for specific missions and operations. The following paragraphs explain the FCCs keys and their functions.

6-9. Although the primary method of operating the FCC is the built-in mouse, the operator can use the **F1 through F10 function keys** (see 1, figure 6-2, page 6-4) to make selections in the software (F11 and F12 are not used). Table 6-1, page 6-4, identifies the assignment of keys.

Chapter 6

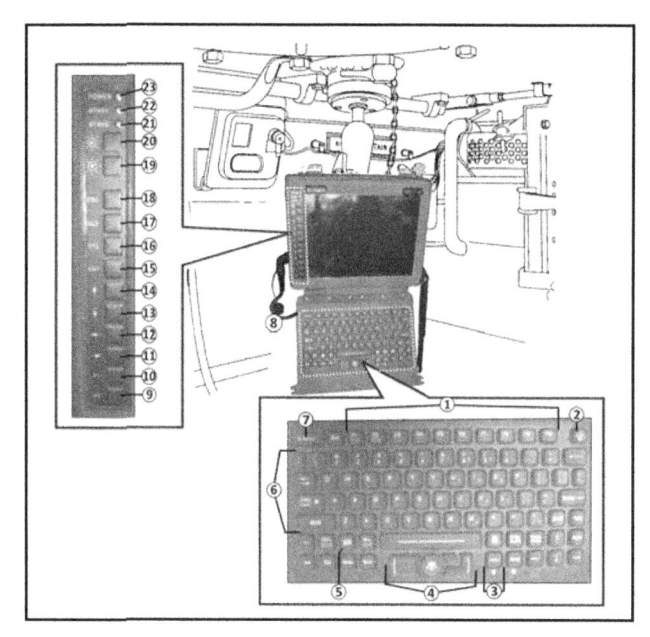

Figure 6-2. Fire control computer

Table 6-1. Function keys

Function Key	Assignment
F1	This Menu
F2	Use All
F3	Undo Changes
F4	Fire Support Coordination Measures
F5	Hipshoot
F6	Final Protective Fire
F7	Boresight
F8	Safety Fan
F9	Checkfire
F10	To Be Determined
F11	Not Used
F12	Not Used

6-10. The **keyboard backlighting control** (see 2, figure 6-2) adjusts the intensity of the light with off, low, and high settings. The **number lock key and indicator** with blue numerals and arithmetic functions can be used as a number pad. Inadvertent use of the number lock (NUM LK) key (see 3, figure 6-2) may result in the inability to perform other desired functions. When activated, the indicator light is illuminated.

Mortar Fire Control System

6-11. The **mouse** (see 4, figure 6-2, page 6-4) allows the user to move the cursor on the screen. The computer **software operating key** (see 5, figure 6-2) is not used by the operator. These alphabetic, numeric, and **special character keys** (see 6, figure 6-2) function as a standard keyboard to compose messages, and are another method to enter data into the system.

6-12. The **blackout key** (see BLACKOUT, 7, figure 6-2) blacks out the screen to guard against enemy detection in a tactical environment. The F1 and F2 **function keys** (see 9 and 10, figure 6-2) make a selection in the software. Mouse or keyboard use is recommended.

6-13. The right, left, down, and up **direction arrow keys** (see 11, 12, 13, and 14, figure 6-2) make a selection in the software. Mouse or keyboard use is recommended. The **enter (ENT) key** (see 15, figure 6-2) brings up a menu of function keys.

6-14. The **control (CTL), alternate (ALT), and escape (ESC) keys** (see 16, 17, and 18, figure 6-2) are not used in this application. The screen **brightness intensity buttons** (see 19 and 20, figure 6-2) decrease and increase the brightness of the LCD screen.

6-15. The **battery 1 (BTRY1) and battery 2 (BTRY2) indicators** (see 21 and 22, figure 6-2) illuminate with a green light when the capacity of the respective battery is 50 to 100 percent of power, with amber light when capacity is 25 to 50 percent, and with no illumination when capacity drops below 25 percent.

6-16. The **power indicator** (see 23, figure 6-2) illuminates with green light when the computer is being powered with external power (power distribution assembly [PDA], alternating current/direct current [AC/DC] adapter) and with amber light when only battery power is used.

POWER DISTRIBUTION ASSEMBLY

6-17. The power distribution assembly (see figure 6-3) accepts direct current (DC) and alternating current (AC) to power the MFCS components. It filters vehicle power through a DC-to-DC power system that isolates the MFCS components from fluctuations in vehicle power, including vehicle starting. It also provides protection to the MFCS components against reverse polarity and power surges. The PDA provides a 115- to 220-volt AC interface for connection to available line power that supports classroom training, as well as nonfield usage, which preserves the battery charge level.

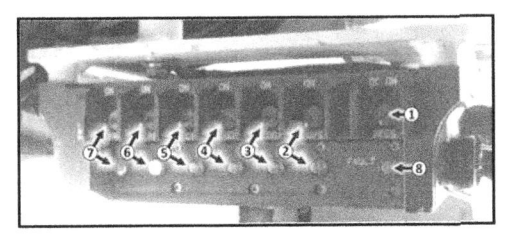

Figure 6-3. Power distribution assembly

6-18. The **power ON switch** (see 1, figure 6-3) turns the PDA on and off only for DC power. The **switch and LED indicator for GPS** (see 2, figure 6-3) distributes power to the PLGR or DAGR. The LED turns green when the switch is ON.

6-19. The **switch and LED indicator for the driver's display** (see 3, figure 6-3) distributes power to the driver's display (DD) only on the M1064. The LED turns green when the switch is ON.

Chapter 6

6-20. The **switch and LED indicator for the gunner's display** (see 4, figure 6-3, page 6-5) distributes power to the gunner's display (GD) on the M1064 only. The LED turns green when the switch is ON. The **switch and LED indicator for the commander's interface** (see 5, figure 6-3) distributes power to the FCC. The LED turns green when the switch is ON. The **switch and LED indicator for the pointing device** (see 6, figure 6-3) distributes power to the pointing device on the M1064 only. The LED turns green when the switch is ON.

6-21. The **switch and LED indicator for the printer** (see 7) distributes power to the printer (PRN) or, in an emergency, the PRN port can be used to power the pointing device. The LED turns green when the switch is ON. This switch is not used in MFCS application except in an emergency situation. The **FAULT LED indicator** (see 8, figure 6-3) illuminates with amber light and, possibly, flickers if the PDA malfunctions.

Pointing Device

6-22. The pointing device (see figure 6-4) is mounted on the mortar tube and aligns the mortar. It can maintain alignment and accuracy within three mils of azimuth and one mil of elevation in all conditions. It provides pointing and positional performance at an operational range of 80 degrees south to 84 degrees north latitude.

6-23. An inertial measurement unit (IMU) provides the weapon with absolute knowledge of vehicle position and mortar barrel azimuth and elevation. The IMU can determine the orientation of the mortar barrel without the need for survey control points, aiming circles, or aiming posts, allowing the mortar platoon to emplace at any location at any time. To maintain a high degree of accuracy, the IMU incorporates information from the GPS and a vehicle motion sensor (VMS). The design of the pointing device allows for the loss of the GPS, the VMS, or both devices without substantial degradation of overall performance.

Figure 6-4. Pointing device

Gunner's Display

6-24. The gunner's display (see figure 6-5) provides the gunner with the necessary information to aim and fire the mortar by displaying deflection and elevation commands by using arrow indicators. It also displays check fire and CFF commands. It is mounted to the left center bipod leg. Function keys are used to start various displays that cover the gunner's functional needs for information, status, and reporting.

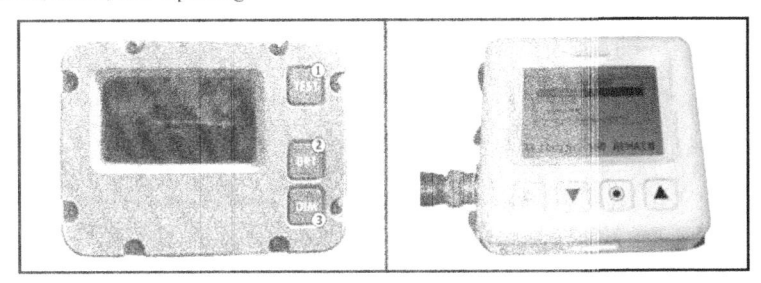

Figure 6-5. Gunner's display

6-25. The **TEST key** (see 1, figure 6-5) initiates the internal built-in test (BIT) display. The **BRT and DIM keys** (see 2 and 3, figure 6-5) increase and decrease the brightness of the LCD. The **locking clamp** (see 4, figure 6-5) is used to allow the gunner to adjust the position of the display.

Driver's Display

6-26. Located within the driver's vision, the driver's display (see figure 6-6) provides the driver with the steering directions and compass orientation to move the vehicle to the next firing location or waypoint. It also provides information to correctly orient the vehicle at the next emplacement. Information is displayed graphically for steering directions and compass orientation, and numerically for distance and heading.

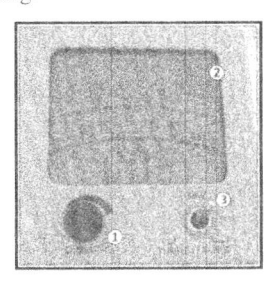

Figure 6-6. Driver's display

6-27. The **dimmer knob** (see 1, figure 6-6) controls the brightness of the LCD. The **liquid crystal display** (see 2, figure 6-6) displays directions to the next position. Blinking of the LCD indicates a CFF. With the **toggle switch** (see 3, figure 6-6) turned to navigation (NAV), the LCD displays directional instructions. The TRK and TURRET/TARGET toggle positions are not used.

6-28. The **vehicle motion sensor** (see figure 6-7, page 6-8) provides the pointing device with vehicle velocity data to reduce the vertical position error and improve location accuracy. Rotation of the drive wheels (that generates the motion of the vehicle) creates a pulse, which represents a

Chapter 6

forward or backward motion of the vehicle. The vehicle motion sensor is mounted inside the engine compartment in front of the driver's cab on the M1064 carrier.

Figure 6-7. Vehicle motion sensor

MORTAR FIRE CONTROL SYSTEM-DISMOUNTED

6-29. The M150 MFCS-Dismounted adds fire control to the M120 trailer mortar system. The M150 (gun) and M151 (FDC) are similar to their heavy counterparts, the M95/M96 MFCS-Heavy. Both systems provide a digitally-integrated fire control system with GPS position, weapon pointing, and ballistic calculation. The primary differences between MFCS-Dismounted and MFCS-Heavy are the base vehicle platforms, power requirements, and component portability.

6-30. The M150 integrates with the M1101 trailer and the M326 MSK. Together, these systems greatly enhance the responsiveness, accuracy, effectiveness, and mobility of the 120-mm towed mortar system. (Refer to most updated TMs.)

6-31. The M150 has some key features and benefits. One key feature of the M150 is it easily integrates into the M1101 trailer with M326. Another feature is that it maximizes the reuse of existing MFCS components and software. The user benefits of the M150 include increased responsiveness, shoot-and-scoot capability, and improved accuracy. The gunner's display is mounted on the bipod, the pointing device mounting assembly (PDMA) is mounted directly onto the cannon of the M120. (See figure 6-8.)

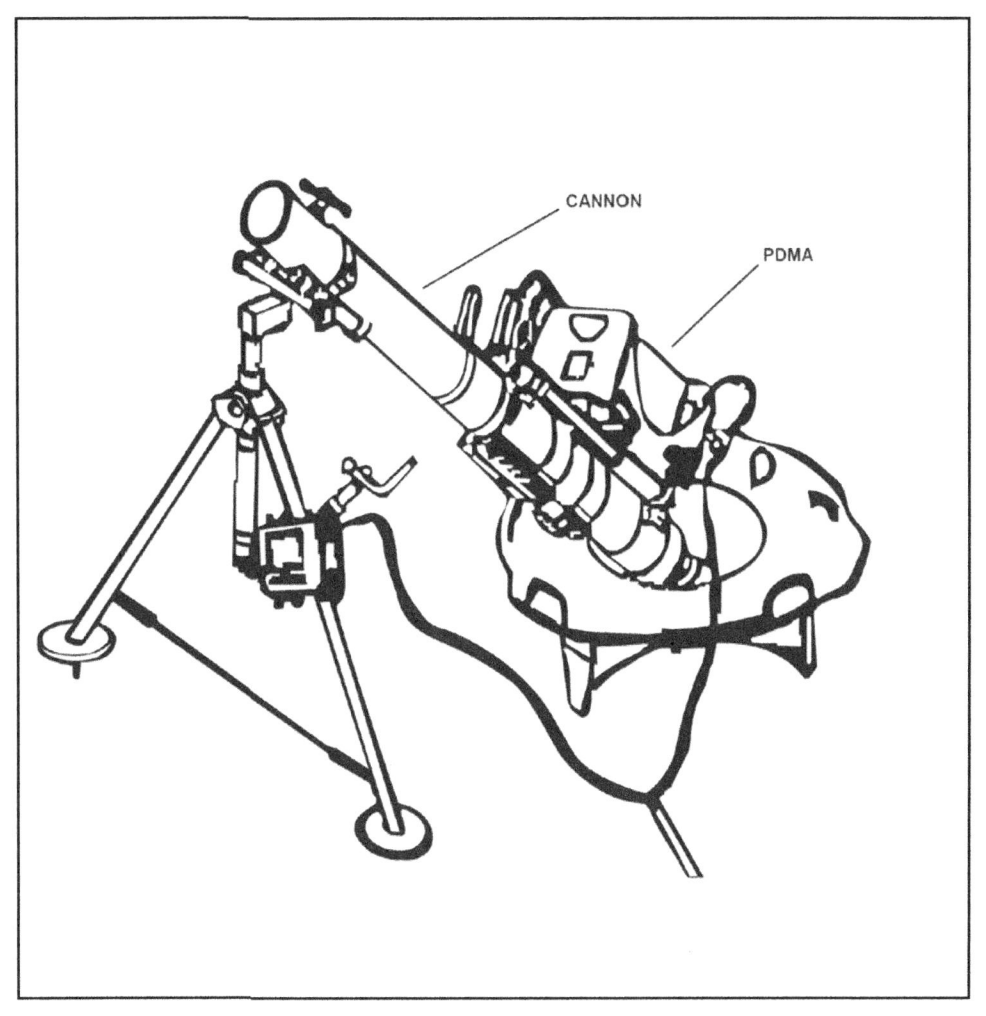

Figure 6-8. M120 with M150 mounted

6-32. The MFCS-Dismounted is an automated fire control system designed to provide improvements in command and control of mortar fires and the speed of employment, accuracy, and survivability of mortars. The vehicle and ground-mounted MFCS-Dismounted components work together to compute targeting solutions, direct the movement of vehicles into firing positions, allow real-time orientation, and present gun orders to the gunner.

6-33. The FCC microprocessor (see figure 6-9) is controlled by a software operating system, manages computer activities, performs computations, and controls the interface with peripheral and external devices. The FCC operator enters data at the touch screen and composes messages using the LCD. Completed messages are then transmitted digitally by radio, two-wire, or local area network (LAN). Each mortar squad also has the capability to compute its own fire missions and can assume the role of a gun or FDC, should the FDC become nonfunctional.

Chapter 6

6-34. System accuracy is increased through the use of a DAGR, an on-board azimuth reference for the gun, and digital meteorological data updates. The MFCS-Dismounted provides the capability to have self-surveying mortars, digital CFF exchange, and automated ballistic solutions. The Dismounted 120-mm Mortar Fire Control System (D120 MFCS) (see figure 6-9) provides a complete, fully-integrated, digital fire control system that can be installed on M326 MSK-equipped trailers (refer to chapter 5 of this publication), mounted in a HMMWV, or operated on the ground.

6-35. The D120 MFCS is capable of sending and receiving digital messages across the battlefield. Navigation instructions are calculated and sent to the FCC through inputs from the pointing device and DAGR.

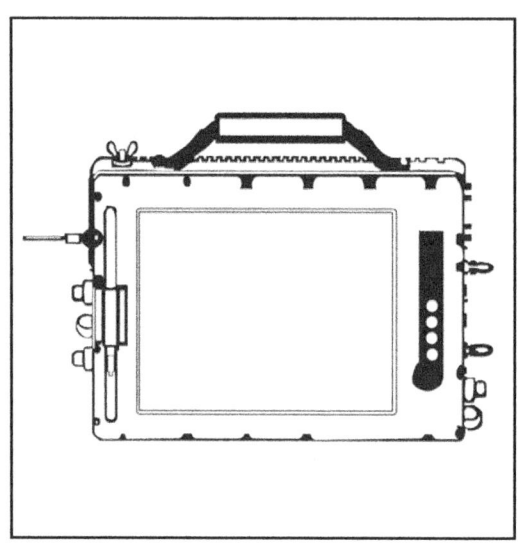

Figure 6-9. Fire control computer

6-36. The major components for the MFCS-D M150 includes—
- Fire control computer.
- Enhanced power distribution assembly (EPDA). (See figure 6-10.)
- Radios (organic).

6-37. However, the MFCS-D M151 has additional features, including—
- Fire control computer.
- EPDA.
- Radios (organic).
- Defense Advanced Global Positioning System Receiver.
- Pointing device with mount assembly dynamic quick release.
- Gunner's display.
- Portable universal battery supply (PUBS).
- Electronic rack.
- Cables, which include the following:
    - The 9W6 that is used to convert the gun to a gun/FDC.

- The 9W10 that is used to convert from a gun to a gun/FDC for the Advanced System Improvement Program (ASIP) B radio.
- The 9W9 that is used to connect to the FBCB2.

FIRE CONTROL COMPUTER

6-38. The FCC is an essential component of the MFCS-D M150 and MFCS-D 151. It manages the information between the gun, FDC, battalion, and other digital subscribers. One of its primary uses is to compute technical fire control solutions while providing an interface between other D120 MFCS-D components and the radio net.

6-39. The FCC serves as a hub for mortar firing operations as it has the storage capacity for target data including fire support control measures. The FCC conducts registration missions and applies all registration corrections automatically. It also has multiple communications and power options that allow it to be used in most any combat or training environment.

ENHANCED POWER DISTRIBUTION ASSEMBLY

6-40. The EPDA (see figure 6-10):
- Receives DC power from external source.
- Provides conditioned DC power to the D120 MFCS components.
- Internal circuit breakers provide protection for each D120 MFCS component.
- The LED indicators monitor power output for each line replaceable unit.
- Provides a communication hub between line replaceable units.
- Monitors EPDA internal serial hub. If a fault is detected, the COMMO fault LED will illuminate.

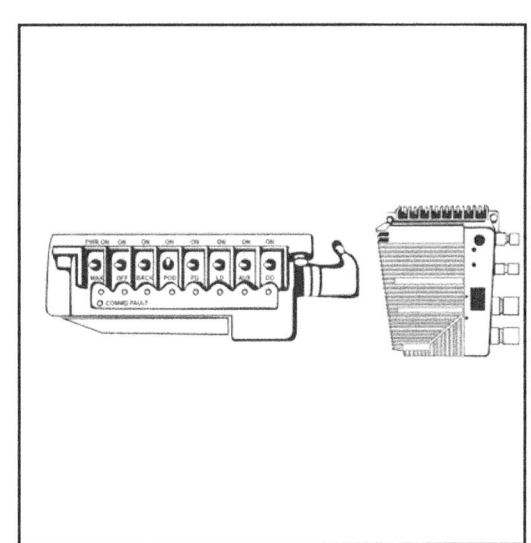

Figure 6-10. Enhanced power distribution assembly

6-41. The radios used with the MFCS-Dismounted have the following characteristics:
- Operates with two ASIP radio sets.

Chapter 6

- One ASIP radio is for the exclusive use for the platoon digital network radio.
- The ASIP radio can be stored in electronics-rack or prime mover.
- The other ASIP, which operates on the digital fire support network.

6-42. Pointing device mounting assembly (PDMA)/Tactical Advanced Land Inertial Navigator (TALIN) (see figure 6-11) have the following characteristics:

- The pointing device is secured in the pointing device quick release assembly. The pointing device quick release assembly is mounted on the mortar tube during firing operations and stowed on the right side of the trailer during mobility mode.
- An inertial measurement unit (IMU) is employed to provide the weapon with absolute knowledge of vehicle position and mortar barrel azimuth and elevation.
- Aligns M120A1 mortar and maintains alignment accuracy within four mils azimuth and three mils elevation in all conditions.
- Robust design of the pointing device allows for the loss of the GPS, without substantial degradation of overall performance.
- The pointing device quick release assembly isolates the pointing device from shock, vibration, and temperature during firing of the gun.

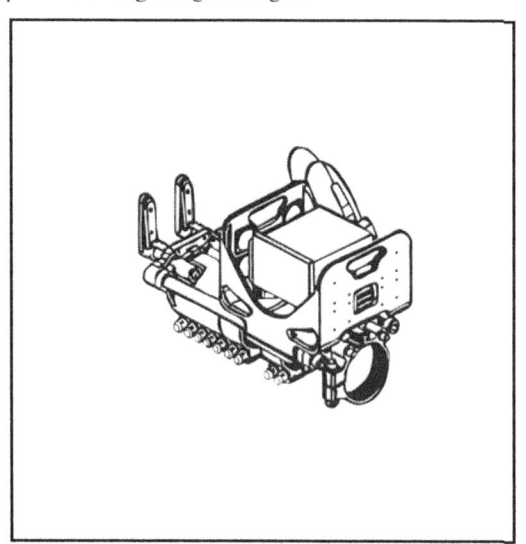

Figure 6-11. Pointing device mounting assembly

6-43. The GD (see figure 6-12) has the following characteristics:

- Mounted to the left bipod leg, the GD provides the information that is necessary for the gunner to aim and fire the mortar in a graphical and textual format.
- Programmable function keys are used to invoke various displays that encompass the gunner's functional needs for information, status, and reporting.

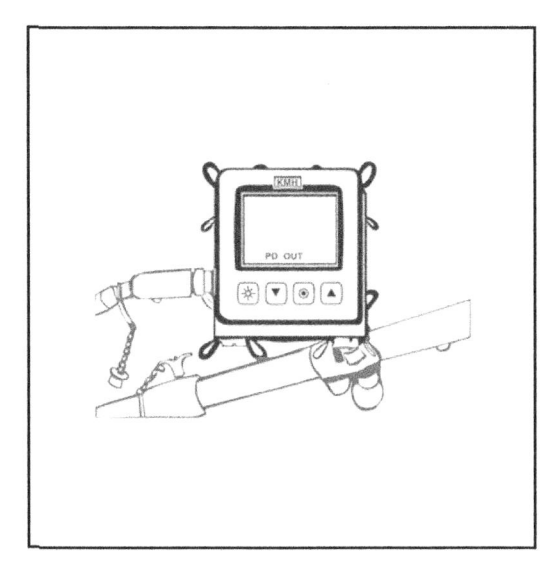

Figure 6-12. Gunner's display

6-44. The DAGR has the following characteristics:
- It is interfaced to the MFCS-Dismounted to provide the MFCS-Dismounted with knowledge of the position and assist the pointing device in maintaining accuracy.
- It serves as a back-up navigation system, if the pointing device becomes inoperative.

6-45. The PUBS, (see figure 6-13, page 6-14) has the following characteristics:
- Is mounted in the electronic rack.
- Provides primary power source for D120MFCS system for about eight hours.
- Rechargeable batteries receive a continuous charge from the prime mover either directly or through the M326 MSK junction box.
- Monitors the output for short-circuit occurrences and provides an audio warning for low battery status.
- Reserve battery provides power for about one hour of operation.

Chapter 6

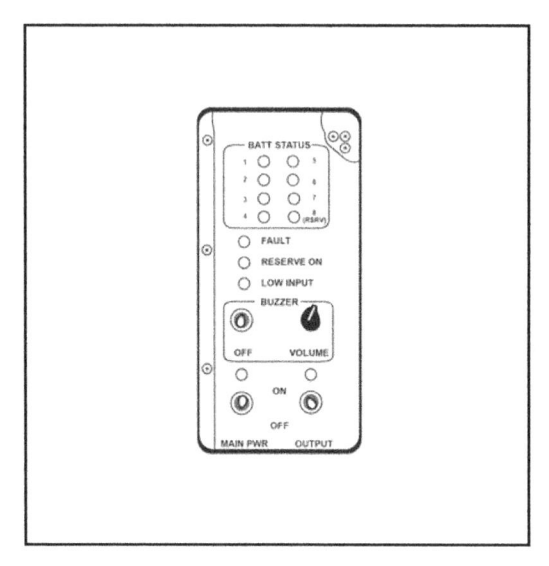

Figure 6-13. Portable universal battery supply

6-46. The electronics rack (ER) (see figure 6-14) has the following characteristics:
- Electronics rack is stowed on the right side of the trailer.
- The DAGR, ASIP, PUBS, EPDA, FCC, GD, and interconnecting cabling and mounting hardware for all onboard electronics are stowed in the ER.
- May be removed from trailer during firing.

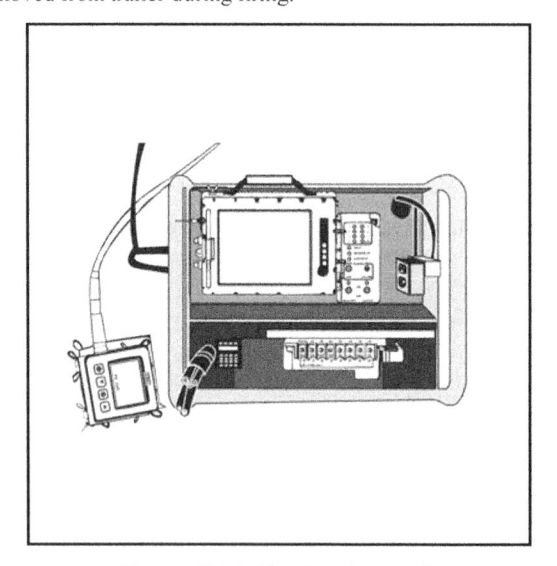

Figure 6-14. Electronics rack

MORTER FIRE CONTROL SYSTEM OPERATIONS

6-47. This section explains how the following MFCS functions work: the pointing device, gun navigation and emplacement, fire command phases, and FPFs.

6-48. The **pointing device** is mounted on the mortar tube in the mortar carrier and aligns the mortar. It determines the correct gun alignment without survey control points, aiming circles, or aiming posts so that the gunner can quickly adjust the mortar to the proper elevation and azimuth to accurately fire the cannon. Throughout the vehicle movement and gun emplacement, the MFCS receives, calculates, and displays instructions to the driver, gunner, and squad leader.

6-49. It can maintain alignment and accuracy within three mils of azimuth and one mil of elevation in all conditions, and provides pointing and positional performance throughout most of the world. The system uses information from a GPS and a vehicle motion sensor (VMS). The design of the pointing device allows for the loss of the GPS, the VMS, or both devices without substantial degradation of overall performance. Prior to a fire mission, the mortar squad ensures that the pointing device is functioning and that the mortar is boresighted properly.

6-50. The **pointing device status screen** (see figure 6-15) shows the status of the pointing device, PLGR, and VMS. This screen displays errors, warnings, and status messages for the pointing device. It also displays the pointing device update position and permits the manual input of position data.

CAUTION

Errors displayed on the status screen require immediate attention to continue the operation.

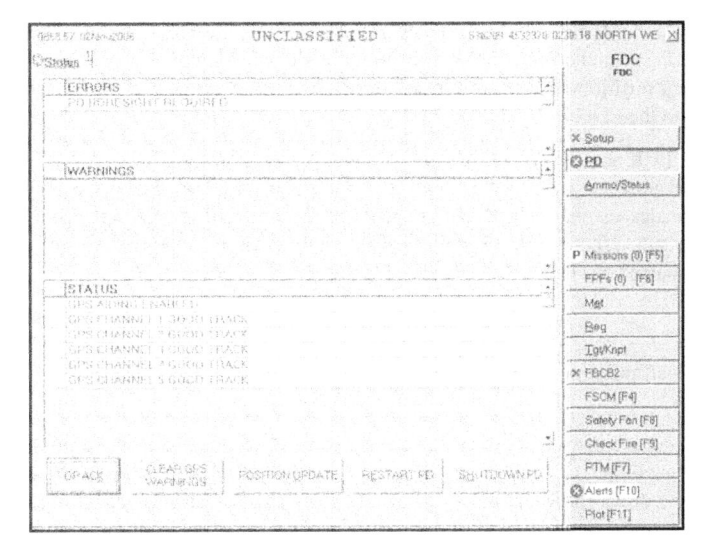

Figure 6-15. Pointing device status screen

Chapter 6

6-51. When the pointing device does not align automatically using PLGR data, it must be updated manually. If the pointing device has not aligned automatically, the words, PD ALIGNING WAIT 600, remain in the upper right corner of the CI screen, and the numbers will not count down automatically.

> **CAUTION**
> To prevent alignment errors, keep movements inside the vehicle to a minimum.

6-52. To manually update the pointing device:

- Click the POSITION UPDATE button. Highlight the information to be changed or click CLEAR POSITION to clear all information in the POSITION UPDATE fields.

- To manually update the pointing device position information, enter the correct data into the EASTING, NORTHING, ALT, ZONE, DATUM, and HEMISPHERE in PD UPDATE POSITION fields. To display or use the last recorded pointing device position, click the SETUP button and select the POSITION tab.

- Click SEND UPDATE. The upper right corner of the screen changes to either ALIGNING: ___SECONDS REMAIN or POS UPDATE IN PROGRESS, depending on whether the pointing device is in need of alignment.

POINTING DEVICE DATA PRECEDENCE RULES

6-53. The pointing device has a set of rules to determine whether to use the manually input locations or the locations generated by the GPS. If the pointing device is aligning using data from a manual position update and the PLGR starts to acquire accurate satellite data, the pointing device will begin to use the PLGR data to complete alignment and, within two minutes, the GPS AIDING OFF box becomes unchecked. When the pointing device alignment is complete, the pointing device's current position is displayed in the upper right-hand corner of the CI screen and is sent automatically to the FDC.

6-54. If the PLGR acquires accurate satellite data after the pointing device has been aligned using data from a manual position update, the GPS AIDING OFF box becomes unchecked and the current coordinates displayed in the upper right-hand corner of the CI screen will begin to drift toward PLGR data. When the drifting is stabilized, the position shown is not automatically sent to the FDC or updated in THIS UNIT'S POSITION or STATUS. The operator must click SEND POSITION UPDATE TO PD. If the two positions do not agree, click SEND POSITION UPDATE TO PD again.

6-55. The restart pointing device button restarts the system after it is shut down. It takes about 10 minutes to restart.

6-56. Following the proper pointing device shutdown procedures ensures a quicker alignment at the next startup, if the vehicle/system has not been moved. To shut down the pointing device without shutting down the computer, press the PERFORM PD SHUTDOWN button. This allows the system to shut down and save its position. It also enables the system to boot up in about two minutes.

Mortar Fire Control System

POINTING DEVICE BORESIGHT SCREEN

6-57. The pointing device boresight screen (see figure 6-16) allows the operator to compensate for mechanical alignment errors and must be completed before using the pointing device. Before beginning, ensure that the power is on and that the pointing device is aligned. The fields at the top left corner display the azimuth, elevation, and roll of the current azimuth correction. Then follow these procedures:

Note. The gunner verifies that the mortar is laid on the DAP at an azimuth of 800 mils, and then clicks the ON DAP AT 800 MILS button.

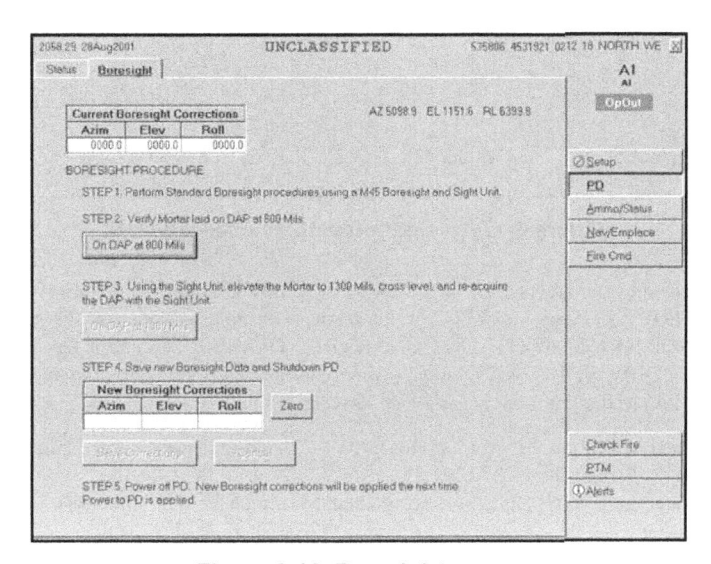

Figure 6-16. Boresight screen

NAVIGATION AND EMPLACEMENT

6-58. The squad leader receives movement orders from the FDC via voice radio or priority text message and inputs the destination into the CI. The MFCS then translates the destination into directions displayed on the driver's display and the CI. These procedures describe how to use waypoints and fire areas for navigation. A waypoint is any nonfiring destination the gun vehicle is required to go (for example, refueling). An FP is any location planned as a position to fire the mortar. (Refer to TM 9-1220-248-10 includes detailed procedures for the emplacement of the M1064A3 gun carrier using the MFCS.)

6-59. Upon receipt of a movement order from the FDC, the squad leader selects the NAV/EMPLACE button in the control button area and clicks the NAV tab to display the navigation/emplacement screen. (See figure 6-17, page 6-18.)

Chapter 6

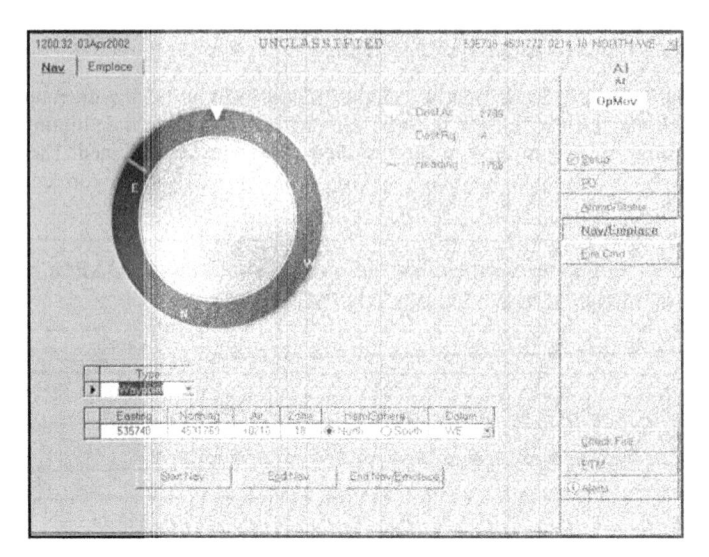

Figure 6-17. Navigation/emplacement screen

6-60. The squad leader enters TYPE by clicking the down arrow and choosing WAYPOINT from choices of FIRE AREA or WAYPOINT, and enters or selects data for the following fields: EASTING, NORTHING, ALTITUDE (ALT), ZONE, HEMISPHERE, and DATUM. The squad leader then clicks the START NAV button, which automatically transmits the status operationally moving (OpMov) to the FDC and activates the driver's display.

6-61. After activation, the driver's display shows the driving direction and distance (see figure 6-18). The driver's display shows STEER TO arrows indicating the direction in which to turn. The destination range and current position is updated continuously until the vehicle approaches the specified area.

Figure 6-18. Driver's display showing steering directions, distance, and position

6-62. The squad leader verifies that the gun status is changed to OpMov. The SEND STATUS box also appears, specifying that an OpMov status was sent to the FDC. The CI now includes the destination azimuth (DestAZ), destination range (DestRg) and heading. If the pointing device is aligned, a position update is sent every 1000 meters.

6-63. When the vehicle/gun is within 30 meters or less of the waypoint, the driver's display indicator displays Arrived. (See figure 6-19.) The CI however, may not display Arrived if the CI has been set for a radius other than 30 meters, in which case the squad leader determines when the vehicle/gun has arrived by the destination range of 30 meters or less.

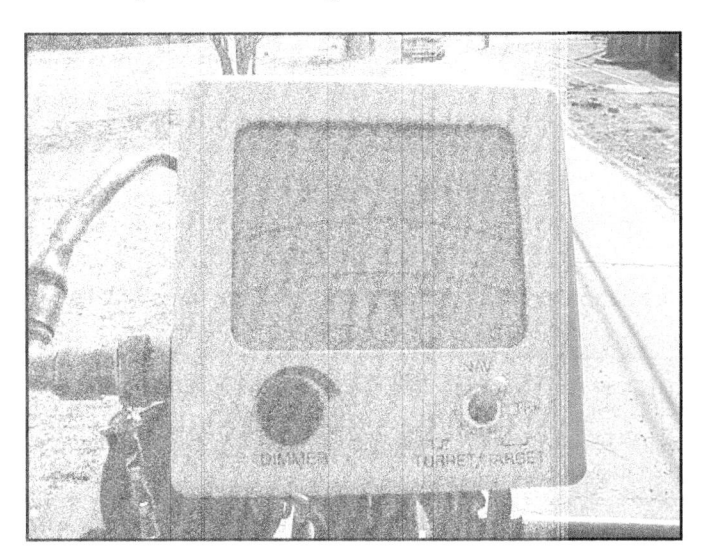

Figure 6-19. Driver's display showing arrival at the waypoint

6-64. The driver continues to maneuver the vehicle left or right until the driver's display indicator is within a minimum of 20 mils and a maximum of 100 mils of the waypoint.

6-65. When the destination is reached, the squad leader clicks END NAV, which sends the FDC an operationally stationary (OpSta) status and the gun's current location. The SEND STATUS box appears, specifying that the message was transmitted and received by the FDC. (See figure 6-20, page 6-20.) At the waypoint, the squad leader accomplishes the task given or waits for further orders.

Chapter 6

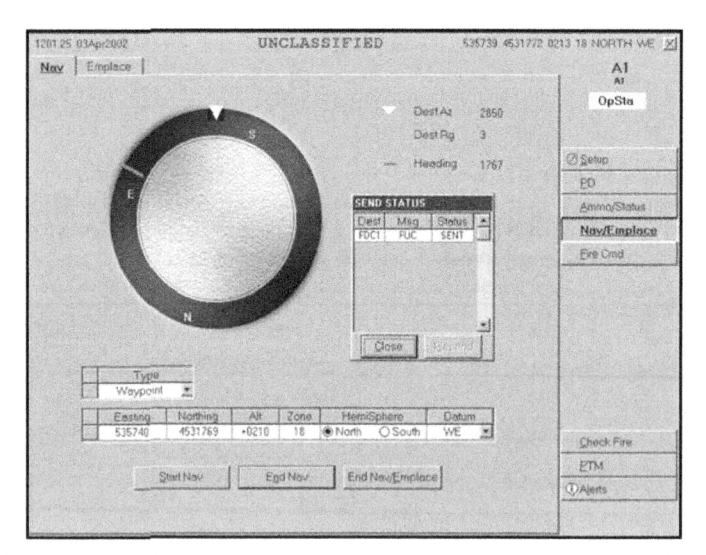

Figure 6-20. Navigation/emplacement screen: message transmittal and reception by the FDC

Note. The current position, which is shown in yellow on the POSITION screen in SETUP and has also been sent to the FDC, is the locked-in current position. The position shown in the upper right-hand corner of the CI may change slightly due to PLGR drift and may differ from the locked-in position.

NAVIGATION TO FIRE AREA

6-66. On receipt of a movement order from the FDC, transmitted through priority text message or voice, the squad leader selects NAV/EMPLACE button in the control button area and clicks the NAV tab. The CI screen shown in figure 6-21 is displayed. In addition to the fields in the WAYPOINT screen, are FIRE AREA RADIUS, AZIMUTH OF FIRE, and ELEVATION fields.

Mortar Fire Control System

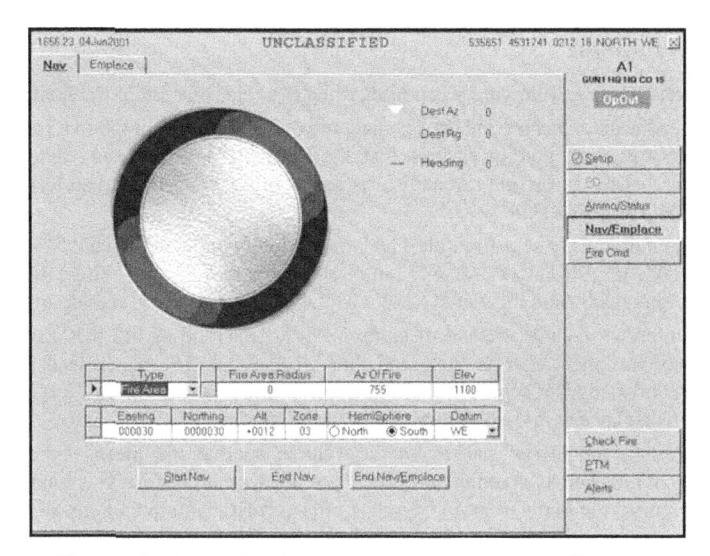

Figure 6-21. Navigation/emplacement screen: fire area

6-67. The squad leader enters TYPE by clicking the down arrow and choosing FIRE AREA from choices of FIRE AREA or WAYPOINT, and then enters the received FIRE AREA RADIUS, AZIMUTH OF FIRE (AZ of fire), and ELEVATION (elev). He also enters the received EASTING, NORTHING, ALTITUDE (Alt), ZONE, HEMISPHERE, and DATUM data into the applicable fields.

6-68. The squad leader clicks START NAV. The driver's display is activated and the driver begins to navigate to the firing position using the steering arrows and distance displayed on the screen.

6-69. On the CI, the gun status is changed to operationally moving (OpMov). The SEND STATUS box appears, specifying that an OpMov status was sent to and received by the FDC. The CI now displays destination azimuth (DestAZ), destination range (DestRg) and heading.

6-70. If the pointing device is aligned, a position update will be sent every 1000 meters. When the vehicle/gun is within 30 meters or less of the fire area, the driver's display indicator shows ARRIVED. If the FIRE AREA RADIUS of 30 meters was entered, the squad leader's CI will also indicate ARRIVED.

Note. If the fire area radius entered in the CI is more than 30 meters, the CI indicates ARRIVED before the driver's display does, because the radius is not changeable in the driver's display.

FIRE COMMAND PHASES

6-71. The FDC uses fire commands to give the mortar sections the information necessary to start, conduct, and cease fire. The fire commands are used for all types of shell and fuze combinations.

6-72. The three phases of the fire command are the initial fire command, subsequent fire command, and the end of mission.

Chapter 6

Initial Fire Command

6-73. The squad uses the following procedures when it receives an initial fire command:

- The squad leader receives and carries out a movement order from the FDC to proceed to his waypoint/fire area. The gunner ensures that the turntable and traversing extension are both centered, and that the traversing mechanism is center of traverse. The squad leader updates and verifies AMMO and STATUS.

- When gun orders are sent from the FDC, the action is indicated by an "I" (immediate) in a red circle beside the FIRE COMMAND button. (See figure 6-23.) The fire command screen is displayed automatically and the words NEW GUN ORDERS appear at the top of the gun order.

- The gun orders include method of control (MOC), method of fire (MOF), target number, phase, rounds, lot, shell, charge, fuze, and time (fuze time). The squad leader reads the gun order and acknowledges receipt by clicking the operationally acknowledge (OpACK) button. The gunner's display and driver's display are now activated.

- If necessary, the driver maneuvers the vehicle so that the mortar tube is pointing a minimum of 20 mils and a maximum of 100 mils of the azimuth of fire. The gunner and assistant gunner manipulate the gun system to bring it to the correct azimuth and elevation displayed on the gunner's display. The gun system is ready to fire when the word LAID appears in both azimuth (AZ) and elevation (EL) of the gunner's display. After firing, the squad leader clicks the SHOT button and if more than one round is shot, the ROUNDS COMPLETE button.

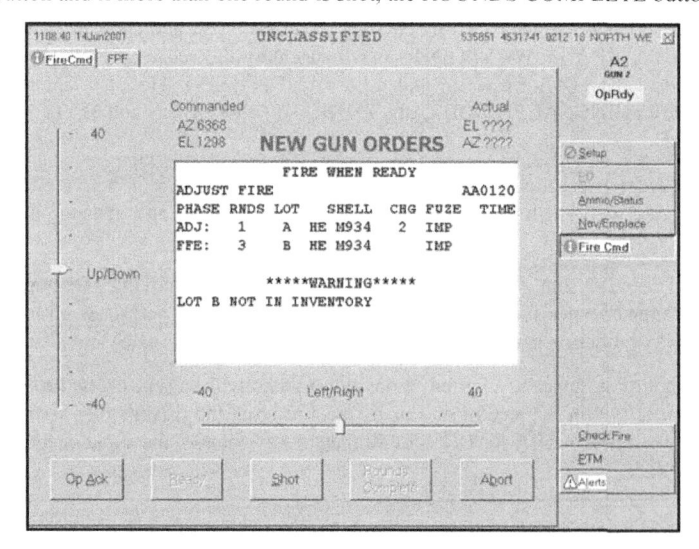

Figure 6-23. Fire command screen

Subsequent Fire Command

6-74. The operator uses the subsequent fire command screen (See figure 6-24.) to receive further commands. The words MORTARS LAID in large blue letters are displayed. If there is a SUBSEQUENT ADJUST, the squad leader receives another gun order.

6-75. The MFCS calculates solutions, and the mortar squad performs the same duties a CFF. The squad continues to respond to commands until the gun is adjusted. Once adjusted, the FDC sends an FFE gun order.

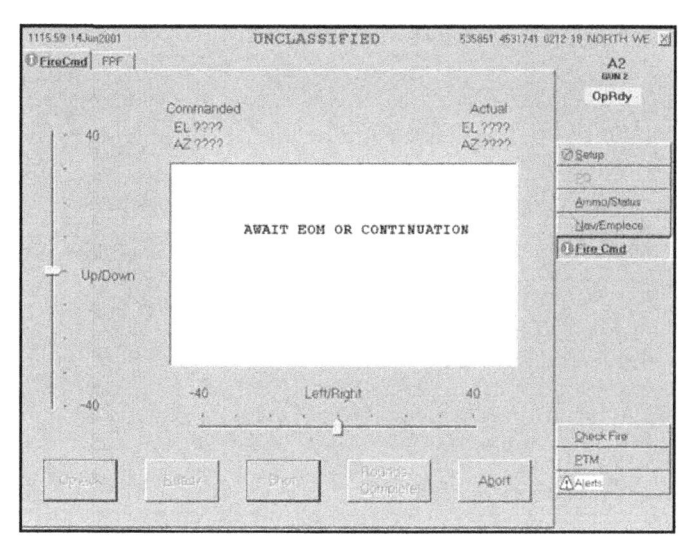

Figure 6-24. Subsequent fire command screen

End of Mission

6-76. Upon receipt of the end of mission command, the END OF MISSION screen displays (see figure 6-25) and the operator confirms the amount of ammunition expended.

6-77. If correct, he clicks YES. When YES is clicked, the NOT IN MISSION screen is displayed. (See figure 6-26, page 6-24.) If incorrect, he clicks NO. When NO is clicked, the AMMO/STATUS screen is displayed and allows a change to the ammunition status.

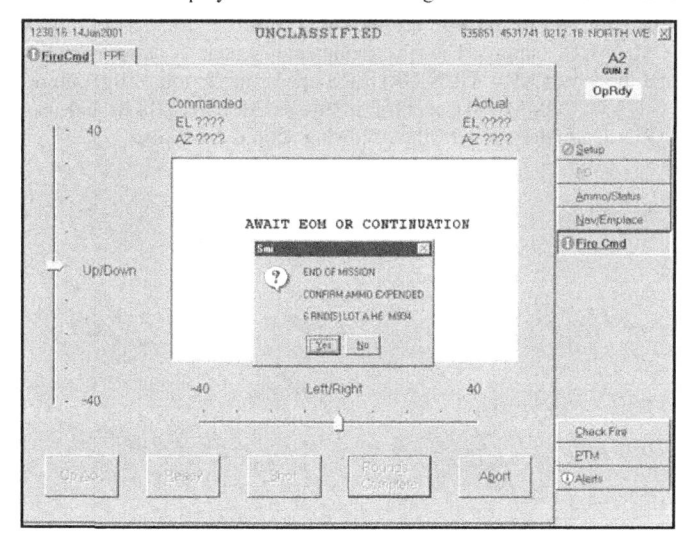

Figure 6-25. End of mission screen

Chapter 6

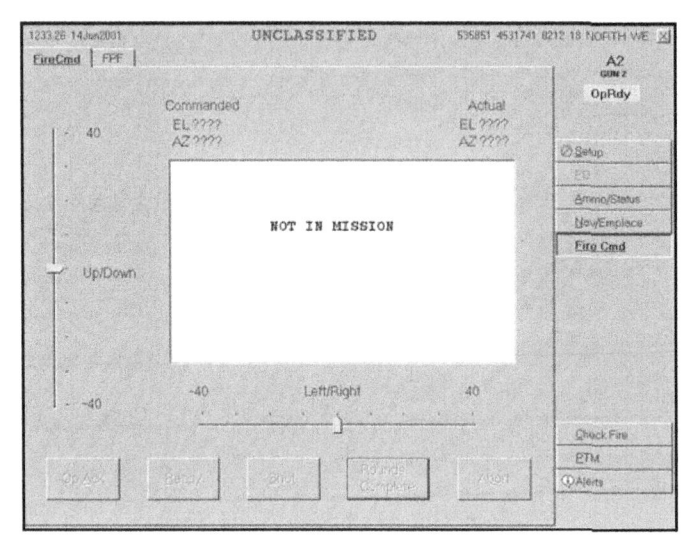

Figure 6-26. Not-in-mission screen

FINAL PROTECTIVE FIRES

6-78. These are the procedures for receiving and carrying out an FPF mission. The MFCS differentiates between an assigned FPF, where the guns are adjusted before an FFE, and a stored FPF, where the guns fire the FPF when ready. The gun MFCS can store only one FPF at a time. If an active mission is in progress and the order to fire the FPF is received, the FPF mission takes precedence.

6-79. Gun orders for an FPF mission are indicated by an "I" (immediate) in a red circle beside the FIRE COMMAND button. The fire command screen is displayed automatically (see figure 6-27) and the words NEW GUN ORDERS appear at the top of the gun order. The method of control is AT MY COMMAND, method of fire is ADJUST FIRE, and method of attack is DANGER CLOSE. Upon these orders, the following actions are taken:

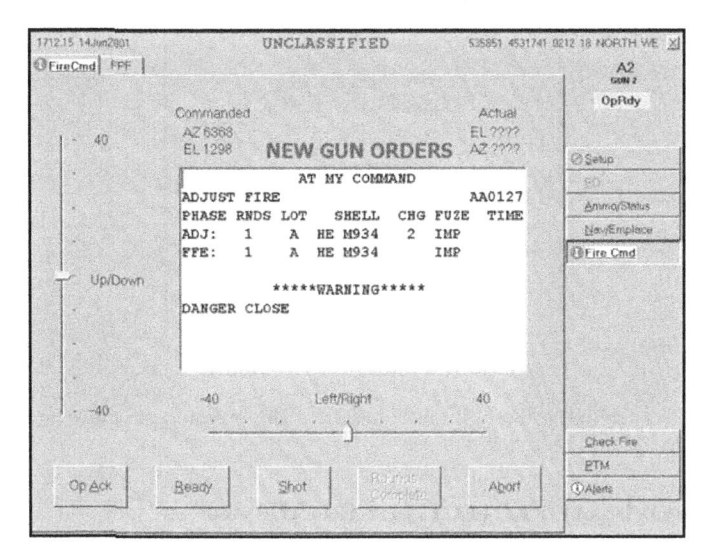

Figure 6-27. Fire command for an assigned final protective fire

- The squad leader reads the gun order and acknowledges receipt. The gunner's display and driver's display are now activated.

- If necessary, the driver maneuvers the vehicle to within a minimum of 20 mils and a maximum of 100 mils of the azimuth of fire. The gunner and assistant gunner manipulate the gun system. The gun system is ready to fire when the word LAID appears in both AZ and EL of the gunner's display. The squad leader sends a READY message to the FDC and waits for a reply to fire.

- After firing the first round, the squad leader sends SHOT. The SEND STATUS box is displayed and then a screen message displays AWAIT END OF MISSION (EOM) or CONTINUATION.

- If there is a SUBSEQUENT ADJUST, the squad leader receives another gun order and acknowledges the order. The mortar squad then follows the same procedures previously described.

- When the gun is adjusted, the FDC sends an FFE gun order. The method of control is AT MY COMMAND.

- Upon receipt of the end of mission command (see figure 6-28, page 6-26), the operator confirms the amount of ammunition expended. (The amount of ammunition expended can be manually changed, if necessary.) When YES is clicked, the message NOT IN MISSION is displayed.

Chapter 6

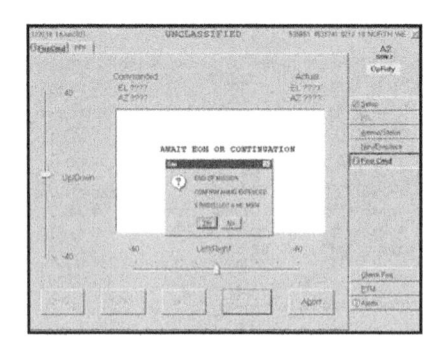

Figure 6-28. End of mission

- The FPF is stored in the FPF mission buffer. The operator can click on the FPF tab to verify that the FPF has been stored.

FIRE A STORED FINAL PROTECTIVE FIRE MISSION

6-80. When the FDC sends the fire the FPF command, the FPF tab is enabled and the FPF screen is displayed. (See figure 6-29.) Then, the following actions are taken:

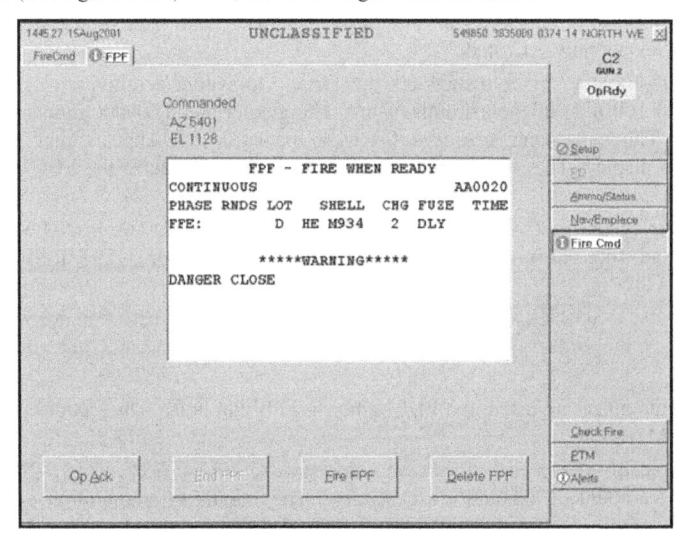

Figure 6-29. Fire the stored final protective fire

- The squad leader reads the FPF and clicks OpACK to acknowledge receipt. The gunner's and driver's displays are activated.
- If necessary, the mortar squad manipulates the gun (as previously described) until the word LAID appears in both AZ and EL of the gunner's display.

Begin firing and continue until the END FPF message is sent. An END OF MISSION message is also sent.

Chapter 7

Fire Without a Fire Direction Center

The use of the target-grid method of fire control may not always be possible or desirable for placing fire on a target. Communications failure, casualties from enemy fire, lack of equipment, or the tactical situation may require that one or more mortars be employed without an FDC. Firing without an FDC greatly increases the vulnerability of the squad and substantially reduces the capabilities of the weapon.

DIRECT LAY AND DIRECT ALIGNMENT

7-1. When the squad is under squad control, the gunner or squad leader observes the target area. Adjustments and fire commands are sent directly to the mortar squad. The two methods used to fire the mortar without an FDC are direct lay and direct alignment.

7-2. In the direct-lay method, the gunner sees the target through the mortar sight. No directional posts, aiming posts, FO, or FDC are used.

7-3. In the direct-alignment method, an FO (squad leader) observes the target near the mortar or the G-T line, observing the fall of the rounds, making corrections relative to the G-T line, and giving gun commands directly to the mortar squad.

Note. Employment of mortars without an FDC is only temporary. The FDC should be established as soon as possible.

7-4. The advantages of operating without an FDC include—
- Speed in engaging a target.
- Better response to commanders.
- Fewer requirements for personnel and equipment.

7-5. The disadvantages of operating without an FDC include—
- Limited movement capability of the forward observer.
- Increased vulnerability to direct and indirect fire.
- Difficulty of massing or shifting fires on all targets within the range of the mortar.
- Necessity of locating the mortar position too far forward where it is subject to enemy fire delivered on the friendly frontlines.
- Greater ammunition resupply problems.

Chapter 7

FIRING DATA

7-6. The direct-lay or direct-alignment methods can be used to lay the mortar for direction. Initial range can be determined by—
- An estimation using the appearance of objects or 100-meter unit of measure methods.
- Map, photographic map, or aerial photograph.
- Mil-relation formula.

7-7. When conducting operations without an FDC, field expedient methods to compute ballistic solutions to fire is by using the ammunition firing table or the ammunition graphical firing scale.

7-8. The **ammunition firing table** provides the Soldier with ballistic data based on standard and nonstandard trajectories. (See the appropriate FT for more information.) It allows the Soldier to chart in on the prescribed lot of ammunition and fuze, and determine the lowest charge to fire based on range to target, corresponding elevation and time of flight.

7-9. The **ammunition graphical firing scale**, commonly known as the whiz wheel, is organic to its specific lot of ammunition and can be found in the ammunition packaging. It allows the Soldier to dial in on range to target, determine lowest charge, time of flight, maximum ordnance, time setting (illumination/registration point [RP]), and corresponding elevation.

7-10. The outer edge of the graphical firing scale contains the range-to-target scale that is graduated in increments of 10 meters and labeled every 100 meters. It can only reaches the maximum range for that specific type of ammunition. The inner scales consist of the corresponding charges, elevations, time of flight, maximum ordnance, and time setting, if applicable. Each whiz wheel contains a guide key located in the center vicinity of the wheel. This key helps decipher the specific ammunition graphical firing scale used to compute ballistic solutions. (See figure 7-1.)

Fire Without a Fire Direction Center

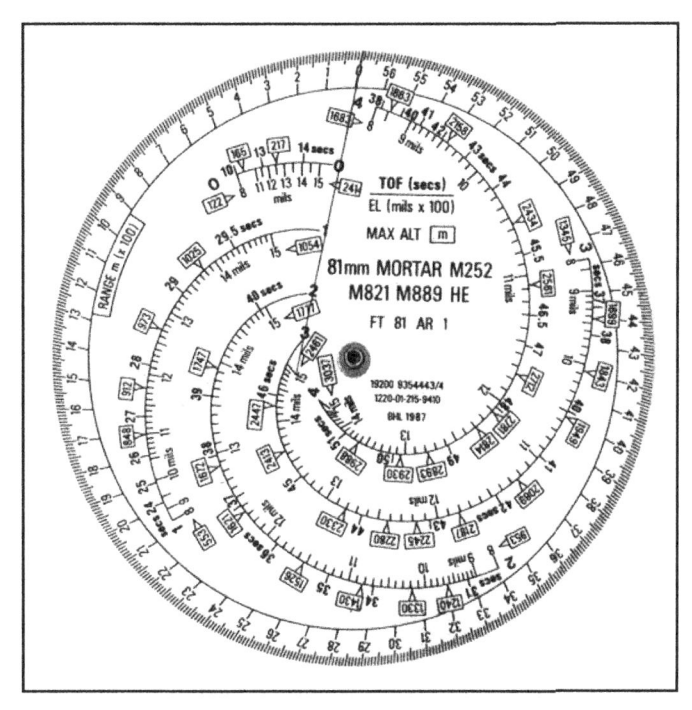

Figure 7-1. Graphical firing scale

DIRECT LAY METHOD

7-11. In the direct-lay method of emplacing a mortar, the gunner sees the target through the mortar sight. No directional or aiming posts, FO, or FDC are used. The firing table should be used to try to obtain a first-round hit.

7-12. If this is not achieved, the firing table should be used to obtain a bracket. Depending on the location of friendly troops to the target, the bracket method, modified ladder method, or creeping method of adjustment apply.

STEP 1. INITIAL FIRING DATA

7-13. The elevation setting and charge selected should be obtained from a firing table. In the absence of a firing table, this information can be determined through the unit SOP or by other expedient techniques such as memorizing charge and elevation for 1000, 2000, and 3000 meters; and firing with the charge and elevation setting closest to the estimated target range. Determine initial range by—

- Estimating.
- Using maps, photographic maps, and so on.
- Intersection.

7-14. Place a 3200-mil deflection on the sight and lay on the center of the target. With the appropriate elevation setting on the sight, center all bubbles by adjusting the lay of the barrel. Take

Chapter 7

appropriate actions to preclude damage to the sight. With established charges, fire the first round, replace the sight, if needed, and observe the burst of the round.

STEP 2: REFERRING THE SIGHT

7-15. Referring the sight centers the vertical line of the sight reticle on the burst. To refer the sight for elevation, take the following actions:
- If the burst is over the target, turn the elevating crank up four, eight, or 16 turns, depending on the gunner's sensing of the round, range to the target, and other possible factors. If the burst is short, turn the elevating crank down four, eight, or 16 turns.
- Turn the sight elevation micrometer knob to center the elevation bubble. If the deflection change requires a bipod displacement, the desired range change is maintained.
- Re-lay the barrel on center of target, centering both bubbles by adjusting the barrel. Fire the second round and observe the burst.

Note. An alternate method of correction for range is to use the firing table.

7-16. To refer the sight for deflection, take the following actions:
- If the burst is to the left or right of the target, refer the sight to the center of the burst by using the traversing crank handle.
- Re-lay the barrel on the center of the target.
- Center both bubbles (elevation and deflection).

STEP 3: BRACKETING THE TARGET

7-17. If the **second round is a line shot and brackets the target**, changing the elevation of the barrel half the number of turns used in step 2 splits the bracket. For example, if the barrel in step 2 was cranked up eight turns, now crank down four and fire the third round.

7-18. If the **second round is not a line shot but does bracket the target**, refer the sight to center of burst and split the bracket by changing the elevation of the barrel half the number of turns used in step 2. Change the sight to the center elevation bubble, and then lay the barrel again on the center of target (centering both bubbles by changing the lay of the barrel). However, if **the second round is not a line shot and does not bracket the target**, repeat step 2 until a bracket is obtained.

STEP 4: FIRE FOR EFFECT

7-19. The appropriate actions of step 3 are repeated until an effect on the target is seen, then mortars fire for effect. After obtaining hits, change the sight to center the elevation bubble and vertical line of the sight reticle on the target, and then record this data. Number the target and retain the number along with the appropriate firing data. The mortar can be taken out of action, moved a short distance, and placed back into action with the mortar able to quickly and accurately attack the recorded target or other close targets.

7-20. If the mortar squad is fired upon during any of the above steps, the mortar can be displaced 75 to 100 meters with minimal effect on the fires, as long as the elevation setting for the last round fired has been recorded or memorized. Once in the new position, use the recorded/memorized data as a starting point and then complete the interrupted step.

Fire Without a Fire Direction Center

DIRECT ALIGNMENT METHOD

7-21. The direct-alignment method uses the aiming posts and FO because the mortar squad cannot direct-lay onto the target visually. The FO/squad leader prepares the initial firing data using the quickest and simplest method available. Initial data consists of a direction of fire and mortar-target range.

7-22. If the **mortar is dismounted**, the squad leader moves to a vantage point where the targets can be seen and places out an aiming post. Then the squad leader directs a member of the mortar squad to place out a second aiming post (to be used as a baseplate assembly stake), aligning it with the first aiming post and target. The mortar squad mounts the mortar at the baseplate assembly stake, places 3200 mils deflection on the sight, and traverses the mortar, aligning the sight on the aiming post placed out by the squad leader. The gunner uses this aiming post as an aiming point or he may place out other aiming posts to be used as aiming points.

7-23. If the **mortar is mounted**, the squad leader moves to a vantage point on a line between the mortar and target. Then the squad leader places out an aiming post (direction stake) on which the gunner lays the mortar with the deflection set at 3200. The gunner uses this aiming post as an aiming point or he may place out other aiming posts to be used as aiming points.

Note. The squad leader can move to a vantage point behind the gun.

7-24. When the tactical situation does not permit alignment of aiming posts and the mortar is placed into action rapidly, the squad leader can use the **natural object method** to establish the mounting azimuth by selecting an object with a clearly defined vertical edge that is situated in the general direction of fire. The squad leader directs the gunner to mount and lay the mortar on the edge of this object with the deflection scale set at 3200 mils. By using the aiming point as a reference point, he can place fire on a target to the right and left by determining the angle in mils between the aiming point and target. The squad leader directs the gunner to place the corresponding deflection on the mortar using the left add/right subtract (LARS) rule.

7-25. The normal method of establishing the initial direction when operating without an FDC is the direct-alignment method. After the initial direction has been established, the forward observer should conduct a registration on the registration point using only the direction stake as a reference point. After registration is completed, a **reference line** should be established by placing out aiming posts on a referred deflection, which then becomes the registration point or base deflection.

FIRE COMMANDS

7-26. When mortars are employed without an FDC, fire commands originate with the squad leader at the observation post.

7-27. Initial fire commands contain the necessary data to lay the mortars and fire the first round. The sequence for transmission of the initial fire command is—
- Mortars to follow.
- Type of projectile (shell) and fuze.
- Mortars to fire.
- Method of fire.
- Deflection.
- Charge.

Chapter 7

- Time setting.
- Elevation.

7-28. The commands used in observed fire procedures without an FDC follow the procedure outlined above with the following exceptions.

7-29. When operating without an FDC, the FO gives the **direction** as a shift from a known point (normally the registration point) in the initial fire command. In subsequent commands, the FO gives the deflection correction from the last round fired. For example, during an adjustment, the FO requests moving the next burst to the right 50 mils. Regardless of the sight setting, the command for deflection is "RIGHT FIVE ZERO." The gunner applies the LARS rule to obtain the deflection to be placed on the sight. The sight has a deflection setting of 20 mils, so the gunner subtracts 50 mils from 20 mils and obtains the new sight setting of 3170 mils. Normally, when the registration has been completed, the aiming posts are placed out on a referred deflection.

7-30. The FO may refer to a firing table, determine the charge and **elevation** (in mils) corresponding to the G-T range, and announce this charge and elevation in his fire command. He may, however, announce the range in meters and have the gunner refer to a firing table to determine the charge and elevation.

7-31. The gunner applies the deflection **correction** to the previous deflection setting, determines the charge and elevation for the given range on the firing table (if required by the FO), announces the charge to the ammunition bearer, sets the elevation on the sight, and then lays for elevation and deflection.

MODIFIED FIRE COMMANDS

7-32. Modified fire commands differ from normal fire commands only because the deflection and elevation changes in subsequent commands are given as turns of the traversing handwheel and elevating crank. The advantages of modified fire commands are speed and simplicity of execution by the gunner. One turn of the traversing handwheel is equal to about 10 mils of deflection for the 60-mm and 81-mm, and five mils for the 120-mm. The unabridged firing tables have a column for the number of turns of the elevation crank to change the range 100 meters.

- When using modified fire commands, deflection and elevation changes are computed to the nearest quarter turn. When the FO anticipates using modified fire commands involving turns of the elevating crank, he announces the range element of the initial fire command as a charge and elevation. This ensures that the FO and gunner are working in the same charge zone, since the number of turns required to move the burst of a round a given distance on the ground can vary considerably between two charge zones.

- The gunner lays the mortar for direction and elevation as given in the initial fire command and does not need to refer to a firing table. Following the initial fire command, the gunner makes no attempt to align the sight on the aiming point or to level the elevation bubble. Corrections are made by taking the turns given in the subsequent commands and keeping the cross-level bubble centered. If the gunner can no longer traverse in the desired direction, he should be able to align the sight on the aiming point, center the traversing bearing, lay again on the aiming point, and resume traverse.

- In computing the number of turns of the elevation crank between two elevations in mils (taken from the firing table), the FO subtracts the smaller elevation from the larger and divides by 10 (for the 60-mm and 81-mm mortars) or by 5 (for the 120-mm mortars). In the ladder and bracketing method of adjustment for range, once the FO has obtained a bracket on the target, he does not need the firing table and continues the adjustment by halving the number of turns of the elevating crank that established the preceding bracket. In the following example, the first round burst is between the FO and target.

> **EXAMPLE (for a 60-mm or 81-mm mortar)**
>
> MODIFIED FIRE COMMANDS:
> RIGHT FOUR TURNS
> DOWN NINE TURNS
> LEFT TWO TURNS
> UP FOUR AND ONE-HALF TURNS
> RIGHT ONE TURN
> DOWN TWO TURNS
> THREE ROUNDS
> UP ONE TURN
>
> NORMAL FIRE COMMANDS:
> RIGHT FOUR ZERO
> ONE SEVEN ZERO ZERO
> LEFT TWO ZERO
> ONE SIX ZERO ZERO
> RIGHT ONE ZERO
> ONE SIX FIVE ZERO
> THREE ROUNDS
> ONE SIX TWO FIVE
>
> In a different example, assume that the first round was fired at a range of 900 meters and burst beyond the target. The FO wants to "DROP ONE ZERO ZERO" (100) for the next round and gives a modified fire command in turns of the elevating crank. Using charge 1, the elevation for the first round at 900 meters is 1275 mils. The elevation for a range of 800 meters is 1316 mils. Subtracting 1275 mils from 1316 mils gives a difference in elevation for the two ranges of 41 mils, or four turns. Therefore, the subsequent command to fire the second round is "UP FOUR TURNS." The second round now bursts short of the target, establishing a bracket. The FO wants to split the bracket and commands, "DOWN TWO TURNS," or half the number of turns that he previously gave to bracket the target. With this command, the FO is splitting a 100-meter bracket and could specify an FFE if he was engaging a tactical area target.

Chapter 7

OBSERVER CORRECTIONS

7-33. In direct alignment, the FO makes corrections differently than when operating with an FDC. All the deviation corrections are made with respect to the G-T line rather than with respect to the observer-target (O-T) line. All deviation corrections are sent in mils or turns of the traversing handwheel. The FO should consistently use one method or the other and not mix the two methods together.

7-34. The FO controls the fire from an observation post, issuing fire commands directly to the mortar squad. The FO may select an observation post close enough to the mortar so that the fire commands are given orally to the mortar squad. When the observation post is not close to the mortar position, the FO uses a telephone or radio to transmit fire commands.

FORWARD OBSERVER WITHIN 100 METERS OF MORTAR POSITION

7-35. The best FO location for rapid fire adjustment on the mortar is at the mortar position. However, the tactical employment of the mortar usually requires the FO to be in a position other than at the mortar.

7-36. If the FO is located within 100 meters of the mortar position, the deviation error he reads in the binoculars can be applied directly to the sight without computation. This is true because the angle between the observer-burst line and O-T line is close to the angle between the mortar-burst line and G-T line. The inherent dispersion of the weapon and the bursting area of the round compensate for any slight difference between these two angles.

7-37. For example, if the FO from a position within 100 meters of the mortar location observes the burst to the left of the target and reads that it is 40 mils left on the mil scale of the binoculars, he orders a correction of RIGHT FOUR ZERO. This correction is sent to the mortar in mils and is not converted to meters. The gunner applies this correction directly to the previous deflection setting using the LARS (left add, right subtract) rule.

FORWARD OBSERVER MORE THAN 100 METERS FROM MORTAR POSITION

7-38. The FO cannot always be located within 100 meters of the mortar position. When the FO cannot locate within 100 meters of the mortar position, he must be within 100 meters of the G-T line. (See figure 7-2.) If he is not, then making the correct adjustments may be difficult. If the FO was attacking targets over a wide frontage, he also would be required to move often.

7-39. Since the angle that exists between the mortar-burst line and G-T line is not equal to the angle that exists between the observer-burst line and O-T line, computations must be made to correct these differences. Because the FO is also the FDC, he must use a correction factor to figure the adjustments from the gun and not from himself.

7-40. The correction factor is a fraction, the numerator of which is the O-T distance and the denominator is the G-T distance (O-T distance/G-T distance). For example, if the FO is halfway between the mortar and target, the correction to be made on the sight is half his deviation spotting; if the mortar is half-way between the FO and the target, the correction is twice his deviation spotting. As other distances give other ratios, a correction factor must be applied to the number of mils spotted before ordering a deflection change. In applying this factor, simplicity and speed are important. The distances used should be to the nearest 100 meters.

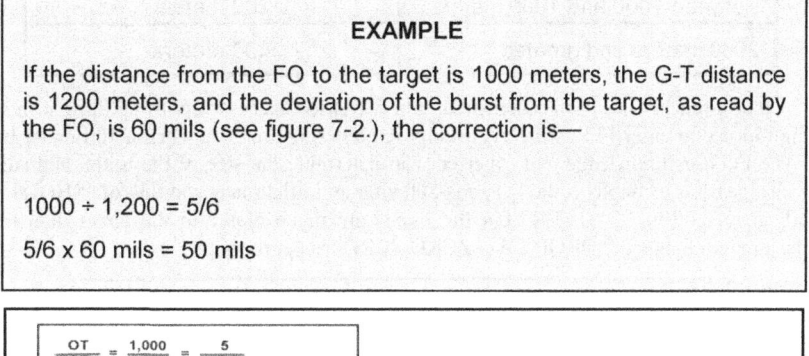

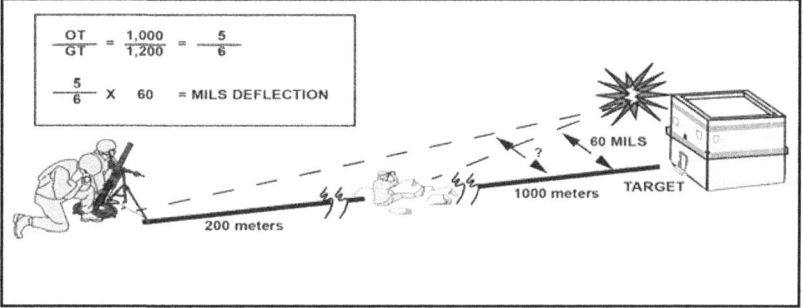

> **EXAMPLE**
>
> If the distance from the FO to the target is 1000 meters, the G-T distance is 1200 meters, and the deviation of the burst from the target, as read by the FO, is 60 mils (see figure 7-2.), the correction is—
>
> 1000 ÷ 1,200 = 5/6
>
> 5/6 x 60 mils = 50 mils

Figure 7-2. Observer more than 100 meters from mortar but within 100 meters of G-T line

ADJUSTMENT METHODS

7-41. When firing without an FDC, the normal procedure for the adjustment of range is the establishment of a bracket along the G-T line. A bracket is established when one group of rounds falls over and one group of rounds falls short of the adjusting point. The FO must establish the bracket early in the adjustment and then successively decrease the size of the bracket until it is appropriate to enter FFE.

BRACKETING METHOD

7-42. When the first definite range spotting is obtained, the FO should make a range correction that is expected to result in a range spotting in the opposite direction. For example, if the first definite range spotting is "SHORT," the FO should add enough to get an "OVER" with the next round. The inexperienced FO should use the guide in table 7-1, page 7-10) to determine the initial range change needed to establish a bracket (see figure 7-3, page 7-10).

Table 7-1. Initial range change

O-T Distance	Minimum Range Change (Add or drop)
Up to 999 meters	100 meters
Between 1000 and 1999 meters	200 meters
2000 meters and greater	400 meters

7-43. Once a bracket has been established, it is decreased successively by splitting it in half until it is appropriate to enter FFE, which is usually requested in area fire when a 100-meter bracket is split. The FO uses knowledge and experience to determine the size of the initial and subsequent range changes. For example, if the FO adds 800 after an initial range spotting of "SHORT" and the second range spotting is "OVER" but the bursts are much closer to the target than the initial rounds, a range change of "DROP TWO ZERO ZERO" is appropriate.

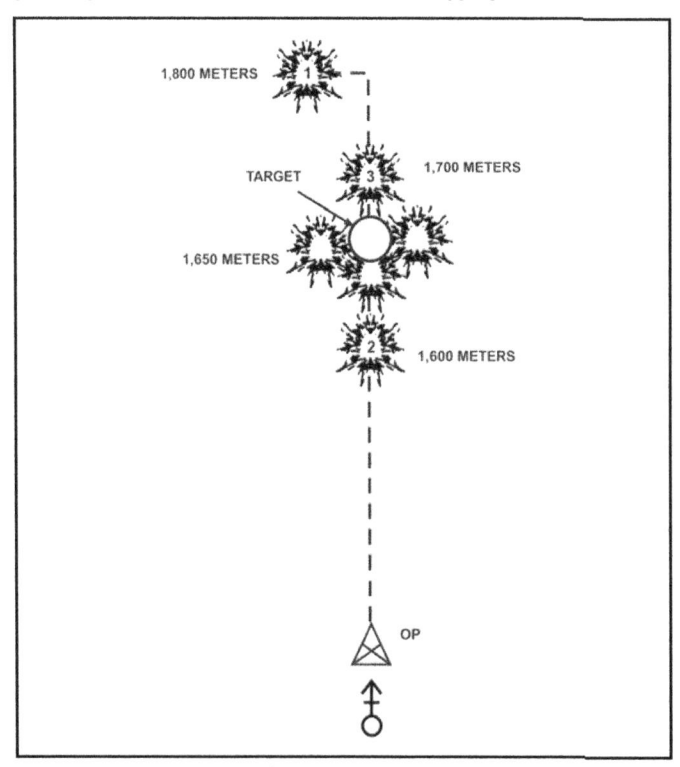

Figure 7-3. Bracketing method

Fire Without a Fire Direction Center

CREEPING METHOD OF ADJUSTMENT

7-44. When a danger close mission is requested, the creeping method of adjustment is used. When the FO requests an adjustment on a target that is within 400 meters of friendly troops, a 200-meter safety factor is added to ensure that the first round does not fall short. When the initial round is spotted, the FO estimates the overage in meters and makes the correction for range by dropping half of the estimated overage.

7-45. Once the FO has given a correction in turns equivalent to a drop in range of 50 meters, OR in the nearest number of half turns equivalent to 50 meters, this same adjustment continues until there is a "RANGE CORRECT," a "TARGET," or a "SHORT" spotting. If, during the adjustment, a round falls short of the target, the FO continues the adjustment using the bracket method of adjustment.

LADDER METHOD OF ADJUSTMENT

7-46. Since surprise is an important factor in placing effective fire on a target, any form of adjustment that reduces the time interval between the burst of the first round for adjustment and FFE is useful. The ladder method of adjustment, a modification of the bracketing method, reduces the time interval and permits FFE to be delivered more rapidly.

7-47. The ladder method may be used by the FO when firing with an FDC. (See figure 7-4.) In the following example, the FO is located within 100 meters of the mortar position.

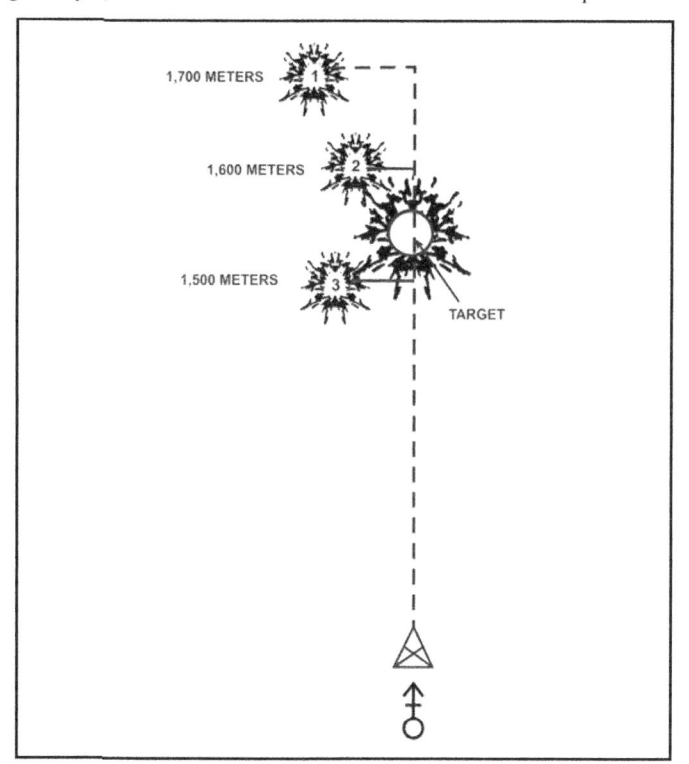

Figure 7-4. Ladder method of fire adjustment

LADDER OF THE METHOD OF ADJUSTMENT

The FO measures the deviation of a target from the registration point as right 30 mils and estimates the G-T range to be 1600 meters. The size of the ladder is based on the minimum range change guide. To obtain a 200-meter ladder, the FO adds 100 meters to this estimated range to establish one range limit for the ladder and subtracts 100 meters from the estimated range to establish the other limit. This should result in a ladder that straddles the target. Three rounds are fired in this sequence: far, middle, and near at 10-second intervals. This helps the FO make a spotting, since no burst is obscured by the dust and smoke from a preceding burst. The FO checks the firing table to obtain the lowest charge and elevation for the far range, 1700 meters (1098); the middle range, 1600 meters (1140); and the near range, 1500 meters (1178) and issues the following initial fire command:

"NUMBER ONE.

HE QUICK.

CHARGE 1.

THREE ROUNDS IN EFFECT.

100-METER LADDER.

FROM REGISTRATION POINT, RIGHT THREE ZERO.

ONE ROUND, ELEVATION, ONE ZERO NINE EIGHT.

ONE ROUND, ELEVATION, ONE ONE FOUR ZERO.

ONE ROUND, ELEVATION, ONE ONE SEVEN EIGHT."

The method-of-fire element in the normal initial fire command for the ladder contains the word ladder. The word ladder tells the gunner that three rounds will be fired as follows: the first at the elevation announced (far range), the second at the announced middle elevation (target range), and the third at the last announced elevation (near range). The gunner indexes the sight for deflection at the base deflection minus 30, while the ammunition bearer prepares the rounds with the designated charge.

The average deviation of all three rounds is left 30 mils. The target is bracketed between the second and third bursts. The squad leader now has a 100-meter bracket of the target between 1500 and 1600 meters, and is ready to FFE. The following subsequent fire command is issued:

"THREE ROUNDS.

RIGHT THREE TURNS.

ELEVATION, ONE ONE SIX ZERO."

ESTABLISHMENT OF A REFERENCE LINE AND SHIFTING FROM THAT LINE

7-48. The normal method of establishing initial direction when operating without an FDC is the direct-alignment method. After initial direction has been established, the FO should conduct a registration using only the direction stake.

7-49. After completing this registration, the FO establishes the G-T line as a reference line. This is accomplished by referring the sight to the desired deflection (usually 3200 or 2800 mils) and then realigning the direction stake on this deflection. The following is an example of the FO located within 100 meters of the mortar position. (See figure 7-5 on page 7-14.)

ESTABLISHMENT OF A REFERENCE LINE

A forward observer has adjusted the fire of the squad on a target. A reference line has been established at 3200 mils on the azimuth scale. The FO observes another target and decides to adjust onto it. He estimates the G-T range to be 1200 meters, determines with binoculars that the target is 60 mils to the right of the first target or registration point, and issues an initial fire command:

"NUMBER ONE.

HE QUICK.

ONE ROUND.

THREE ROUNDS IN EFFECT.

SHIFT FROM RP1, RIGHT FIVE ZERO.

RANGE ONE TWO ZERO ZERO."

The gunner sets the sight with a deflection of 3200 mils and applies the LARS rule (50 mils subtracted from 3200 mils equals a deflection of 3150). Looking at the firing table for HE M821 ammunition, the charge is 1 at elevation 1278. The gunner announces the charge to the ammunition bearer and at the same time, places the elevation on the sight. The gunner lays the mortar and commands, "FIRE."

The FO spots the first round burst as over the target and 25 mils to the left. The squad leader issues a subsequent fire command correcting the deflection using the minimum range change, and decreasing the range 200 meters between himself and the target. In doing so, the FO establishes a bracket and then commands, "RIGHT TWO ZERO (20), RANGE ONE ZERO ZERO ZERO."

The second round bursts between the FO and target and on the G-T line. The deflection is now correct, and a 200-meter bracket has been established. The squad leader's subsequent fire command is, "RANGE, ONE ONE ZERO ZERO." The gunner selects the charge and elevation from the firing table, announces the charge to the ammunition bearer, and when all is ready commands, "FIRE."

Chapter 7

$$\frac{OT}{GT} = \frac{1,000}{1,200} = \frac{5}{6}$$

$$\frac{5}{6} \times 60 = 50 \text{ MILS DEFLECTION}$$

FIRST TARGET

60 MILS

NEW TARGET

200 METERS — 1,000 METERS

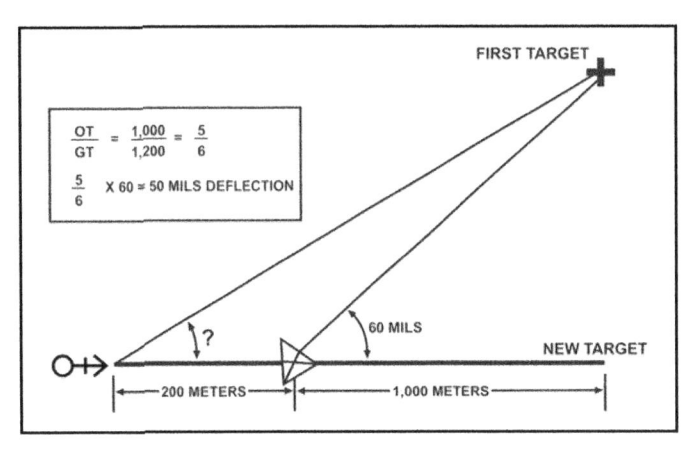

Figure 7-5. Adjustment of fire onto a new target with the observer within 100 meters of the G-T line

---

ESTABLISHMENT OF A REFERENCE LINE (cont.)

This third round bursts beyond the target and on the G-T line. A 100-meter range bracket now has been established. In the next fire command, the squad leader combines the adjustment with FFE:

"THREE ROUNDS.

RANGE, ONE ZERO FIVE ZERO."

The bursting area of these rounds and their normal dispersion cover the target area with casualty-producing fragments. If the FFE fails to cover the target adequately, the squad leader makes any necessary changes in deflection or range and again orders FFE. This adjustment can also be fired using modified fire commands. The initial fire command for this mission then would be:

"NUMBER ONE.

HE QUICK.

ONE ROUND IN ADJUST.

THREE ROUNDS IN EFFECT.

FROM RP.

RIGHT FIVE ZERO.

CHARGE 1.

ELEVATION ONE TWO SEVEN EIGHT."

SPOTTING

7-50. A spotting is the FO's determination of the location of the burst with respect to the target. Spottings are made for the range and the deviation from the G-T line. The FO must be able to visualize the G-T line in order to orient the bursts to it.

7-51. **Definite range spottings** are required to make a proper range adjustment. Any range spotting other than "DOUBTFUL," "LOST," or "UNOBSERVED" is definite. Definite range spottings may include a—
- Burst or group of bursts on or near the G-T line.
- Burst(s) not on the G-T line but located by the forward observer using his knowledge of the terrain or drifting smoke, shadows and wind patterns. Use of these spottings requires caution and good judgment.
- Location of the burst fragmentation pattern on the ground.
- Burst that appears beyond the adjusting point (over).
- Burst that appears between the forward observer and adjusting point (short).
- Round that bursts within the target area (target).
- Burst or center of a group of bursts that is at the proper range (range correct).

7-52. **Doubtful, lost, or unobserved range spottings** cannot be used for adjustments and a bold shift in deviation or range should be made for the next round. These range spottings include:
- Doubtful: a burst that can be observed but cannot be determined as over, short, target, or range correct.
- Lost: a burst whose location cannot be determined by sight or sound.
- Unobserved: a burst not observed but known to have impacted (usually heard).
- Unobserved over or short: a burst not observed but known to be over or short. This spotting provides some information to assist the forward observer in the subsequent adjustment.

7-53. A **deviation spotting** is the angular measurement from the target to the burst. There are three possible deviation spottings:
- Line: a round that impacts on the G-T line
- Left: a round that impacts to the left of the G-T line.
- Right: a round that impacts to the right of the G-T line.

7-54. A round that varies greatly from normal behavior is classified as an **erratic round**.

SQUAD CONDUCT OF FIRE

7-55. Conduct of fire includes all operations in placing effective fire on a target. Examples include the FOs ability to open and adjust fires, the ability to distribute fire on the target, shifting fire from one target to another, and regulating the type and the amount of ammunition to be expended. Quick actions and teamwork are required to efficiently conduct fires. Training increases the squad's ability to fire without an FDC.

7-56. The normal sequence of instruction begins on the training shell range using practice ammunition, and progresses to field training exercises with practice or combat ammunition. Training ammunition allocation is outlined in DA PAM 350-38.

7-57. To ensure maximum efficiency, each squad member is acquainted with the principles of technique of fire for each type of adjustment and FFE. Frequent rotation of duty helps squad

Chapter 7

members to better understand this technique of fire. The designated FO (squad member) is trained in all methods and techniques used in bringing effective fire on a target as quickly as possible.

USE OF SMOKE AND ILLUMINATION

7-58. Smoke is used to obscure the enemy's vision for short periods. Illumination is designed to assist friendly forces with light for night operations.

7-59. The squad leader must be authorized to **employ smoke**. The authority to fire smoke rests with the battlefield commander that the screen will affect. After careful evaluation of the terrain and weather, the forward observer locates a point on the ground where to place the screen. If necessary, the FO adjusts fire to determine the correct location of this point. For a screening mission, splitting a 100-meter bracket normally is sufficient.

7-60. The battalion commander exercises control over the use of Infantry mortar **illuminating rounds** after coordination with adjacent units through the next higher headquarters. The correct relative position of the flare to the target depends upon the wind and terrain. The point of burst is placed to provide the most effective illumination on the target and to ensure that the final travel of the flare is not between the forward observer and target. Adjusting the round directly over the target is unnecessary due to the wide area of illumination.

7-61. If there is a strong wind, the point of burst must be placed upwind from the target so the flare drifts to the target location. The flare should be slightly to one flank of the target and at about the same range. When the target is on the forward slope, the flare is placed on the flank and at a slightly shorter range.

7-62. For adjustment on a prominent target, better visibility is obtained by placing the flare beyond the target to silhouette it and prevent adjustment on the target's shadow. When firing continuous illumination, a strong wind can decrease the time interval between rounds. For maximum illumination, the flare is adjusted to burn out shortly before reaching the ground.

SEARCH OR TRAVERSE

7-62. Mortars use search or traverse when target areas cannot be completely engaged with linear or open sheaves. To increase the depth of a target area, mortars employ searching fire.

7-63. To increase the breadth of a target area, mortars employ traversing fire. Search and traverse may be employed simultaneously to engage a target that is deeper and wider than a linear sheaf.

SEARCHING FIRE

7-63. The forward observer uses searching fire to place effective fire on deep targets. To engage a deep target, the FO adjusts on one end of the target, normally the far end. When anticipating the use of searching fire, the FO announces the range as a charge and elevation. Then, the following actions are taken:

- After adjustment is completed on one end of the target, the FO estimates the range to the other end of the target. Use the firing table to determine the number of turns of the elevating crank needed to change the range 100 meters. If the FO does not have this information, the number of turns can be decided by determining the difference in elevation in mils that exists between the two ranges, and dividing this difference by 5 or 10 to determine the number of turns needed on the elevating crank to cover the target.

Fire Without a Fire Direction Center

- The forward observer must then determine the number of rounds to be fired. Usually five rounds cover an area 100 meters deep, except at long ranges where dispersion is greater. Once the number of rounds to be fired is determined, then the number of intervals between rounds is decided. There always will be one less interval than the number of rounds fired.
- The FO then divides the total number of turns required by the total number of intervals to determine the number of turns the gunner must make between each round, computed to the nearest half turn. (See figure 7-6, page 7-18.)

ADJUSTMENT OF AN 81-MM MORTAR

If the FO has adjusted to the far end of the target and found it to be 1000 meters. The FO estimates the near edge of the target to be 950 meters. By using the firing table, the FO determines to make five turns of the elevating crank to change the range 100 meters. Since the FO only wants to make a 50-meter change, only half of the turns are made, which is 2 1/2 turns. If this information is not available, the FO would determine that there is a 23-mil difference in elevation for the two ranges (elevation 1231 mils for range 1000 meters, and elevation 1254 mils for range 950): dividing by 10, it would require two turns of the elevating crank.

The FO has determined to use two rounds to attack the target. There will be one interval between the two rounds fired. Then, the FO divides the total number of turns required by the number of intervals, and rounds off the answer to the nearest half turn.

2 1/2 ÷ 1 (interval) = 2 1/2 turns between rounds

The FO is now ready to send the subsequent fire command to the gunner. The commands are—

"TWO ROUNDS.

SEARCH UP TWO AND ONE-HALF TURNS.

ELEVATION ONE TWO THREE ONE."

The gunner is told to search in the direction that the barrel moves. For this example, the barrel moves from 1231 mils to 1254 mils of elevation; therefore, the command is "SEARCH UP."

---

17 March 2017   TC 3-22.90   7-17

Chapter 7

```
                    GT/OT LINE
                        |
                        |
    1,000 METERS    ┌───────┐  ELEVATION
                    │       │  1231 MILS
    TARGET   ────►  │       │
                    │       │
    (950 METERS)    │       │  ELEVATION
                    └───────┘  1254 MILS
                        |
                        |              1254 MILS
                        |            - 1231 MILS
                       ≈≈≈            ───────────
                        |             23 MILS DIFFERENCE
                        |                IN ELEVATION
                        |
                       △               23 MILS
                                      ─────────── = 2.3 TURNS
                                      1 INTERVAL
                       ↑
                       ○               2 ½ TURNS
                                      ─────────── = 2 ½ TURNS
                                      1 INTERVAL   BETWEEN
                                                   ROUNDS

                                      MODIFIED FIRE COMMAND
                                         TWO ROUNDS
                                         SEARCH UP 2 ½ TURNS
                                         ELEVATION 1231 MILS
```

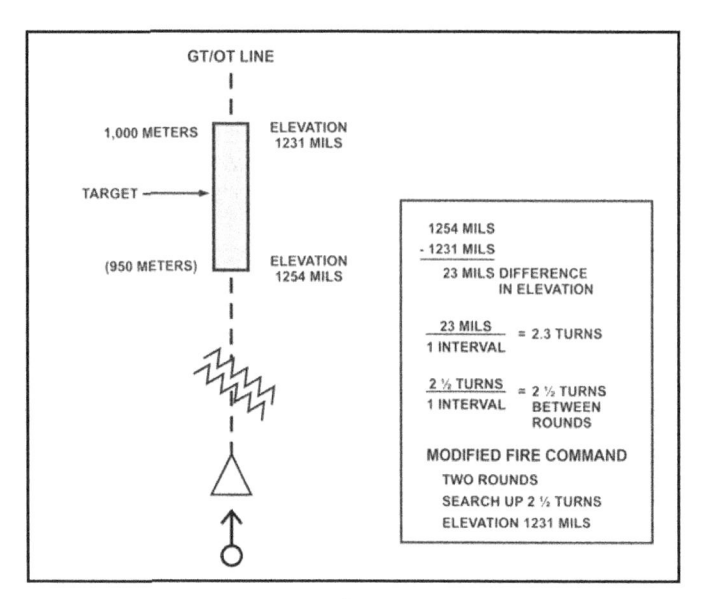

Figure 7-6. Searching fire

TRAVERSING FIRE

7-64. To attack wide targets, the FO must use distributed FFE. In distributed FFE, the gunner fires a specified number of rounds but manipulates the mortar for range or deflection between each round. Distributed fire on wide targets is called traversing fire. To place traversing fire on a target, the forward observer adjusts fire on one end of the target, normally the end nearest to a known point. Then the following actions are taken:

- After adjustment, the FO determines the width of the target in mils by using the mil scale in the binoculars, or by reading an azimuth to each end and subtracting the smaller from the larger. The FO then divides the mil width by 10 for the 60-mm and 81-mm, and 5 for the 120-mm. (Ten and five are the number of mils that one turn of the traversing handwheel moves the mortar.) This determines the number of turns needed to traverse across the target, computed to the nearest half turn.

- In computing the number of rounds, the FO divides the width of the target by the bursting area of the round. The FO then divides the total number of turns by the number of intervals between the rounds to be fired to determine the number of turns between rounds, computed to the nearest half turn. There will always be one less interval than the number of rounds fired in the FFE phase.

- After adjustment and before issuing the subsequent fire command, the FO tells the gunner to prepare to traverse right or left. The gunner traverses the mortar all the way in the direction commanded and then back two turns (four turns on M120/M121) on the traversing handwheel. With the aid of the assistant gunner, the gunner moves the bipod legs until it is approximately re-laid on the direction stake. Using the traversing mechanism, the gunner then completes realigning the mortar and announces, "UP."

- When the mortar is laid, the FO issues the subsequent fire command, announces the number of rounds to be fired, and the amount of manipulation between each round.

7-65. In the example in figure 7-7, the FO is located with the 81-mm mortar squad and measures the width of the target to be 75 mils. Using a map, the FO estimates the range to be 2200 meters. Using the mil-relation formula, the width of the target is determined to be 165 meters. The FO decides to attack the target with seven rounds. There will be six intervals between the seven rounds. Since the target is 75 mils wide, the number of turns is determined to be 7 1/2. To determine the number of turns between rounds, the number of turns is divided by the number of intervals (7 1/2 divided by 6 = 1.07). This is rounded off to the nearest half turn (one turn).

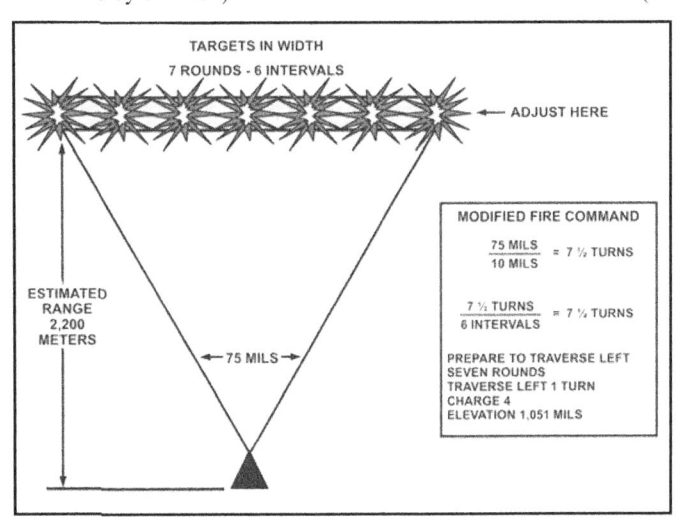

Figure 7-7. Traversing fire

Note. When determining the rounds for FFE, the forward observer applies the following rules:
    1. For the 60-mm and 81-mm mortars, one round for each 30 meters and four rounds for each 100 meters.

    2. For the 120-mm mortar, one round for each 60 meters and two rounds for each 100 meters.

MOVEMENT TO ALTERNATE AND SUPPLEMENTARY POSITIONS

7-66. When time or the situation dictates, the mortar may be moved to alternate and supplementary positions, and registered on the registration point, FPF (in the defense), and as many targets as possible.

This page intentionally left blank

Chapter 8

Gunner's Examination

The gunner's examination tests the proficiency of the gunner in five areas: mounting the mortar, making small deflection change, referring the sight and realigning aiming posts, making large deflection and elevation changes, and reciprocal lay of the mortar. It is also a test of the three qualified assistants the candidate is allowed to choose. The candidate's success in the examination depends mainly on the ability to work harmoniously with these assistants. The examining board must consider this factor and ensure uniformity during the test. Units should administer the gunner's examination at least semiannually to certify squad proficiency.

PREPARATION AND INSTRUCTION

8-1. Preparation for the gunner's examination teaches the Soldier how to properly and accurately perform the gunner's duties. The squad leader is responsible for this preparation.

8-2. In table of organization and equipment units, squad members should be rotated within the squad so that each member can become proficient in all squad positions. Individual test scores should be maintained and consolidated to determine each squad's score. These squad scores then can be compared to build esprit de corps.

Note. Stryker mortar gunner's examination is covered in chapter 8.

8-3. The conditions and requirements for each step of the qualification course is explained and demonstrated. Then, each candidate is given practical work while being constantly supervised by their squad leader to ensure accuracy and speed. Accuracy is stressed from the start and speed is attained through repetition.

8-4. The platoon leader/platoon sergeant monitors the instruction given by the squad leaders within the platoon. Demonstrations usually are given to the entire group. Squads perform practical work under the supervision of the squad leader.

8-5. A Soldier must be proficient in mechanical training, squad drill, and fire commands and their execution before being qualified to take the examination. Without the proper prior training, a Soldier should not be tested.

8-6. The preparatory exercises for the gunner's examination consist of training in those steps found in the qualification course. After sufficient preparatory exercises, candidates are given the gunner's examination. Those failing the examination should be retrained for testing at a later date.

Chapter 8

EXAMINATION

8-7. The examining board consists of one officer and two senior NCOs who are proficient with the weapon. No more than one member is selected from the candidate's organization. The commander who has authority to issue special orders appoints the board. Scores are recorded on DA Form 5964. (See figure 8-1.)

Gunner's Examination

GUNNER'S EXAMINATION SCORECARD - MORTARS

For use of this form see TC 3-22.90, the proponent agency is TRADOC.

1. MORTARS (CHECK ONE)
- Ground-Mounted ☒
- Carrier-Mounted ☐

DATA REQUIRED BY THE PRIVACY ACT OF 1974

AUTHORITY: 10 USC 3012 (g), Executive Order 9397.
PRINCIPAL PURPOSE: To evaluate an individual's training on mortars.
ROUTINE USES: Information evaluates individual proficiency; Service ID# is used for positive identification purposes only.
DISCLOSURE: Disclosure of this information is voluntary. However, individuals not providing information cannot be rated/ scored on a mass basis.

2. ORGANIZATION
HHC 1st BN (MECH) 66th INF

3. DATE (YYYYMMDD)
20170510

4. NAME (Last, First, MI)
Doe, John J.

5. GRADE
E-4

6. CHECK ONE
- 60MM ☐
- 81MM ☒
- 120MM ☐

ITEMS	TRIAL (a)	TIME (b)	TRIAL SCORE (c)	CREDIT SCORE (d)	INITIALS (e)
7. GROUND - Mounting the mortar OR CARRIER - Placing the mortar in firing position from traveling position	1st Trial	005800	9	18	JBO
	2nd Trial	005300	9		
8. ALL - Small deflection and elevation change	1st Trial	001300	8	18	JBO
	2nd Trial	000800	10		
9. ALL - Referring the sight and realigning aiming posts	1st Trial	011000	10	19	JBO
	2nd Trial	011300	9		
10. ALL - Large deflection and elevation change	1st Trial	003800	8	16	JBO
	2nd Trial	004000	8		
11. ALL - Reciprocal laying	1st Trial	004700	9	19	JBO
	2nd Trial	004400	10		

12. TOTAL CREDIT SCORE (Check one)
90

RATING	
90 to 100	EXPERT GUNNER ☒
80 to 89	FIRST-CLASS GUNNER ☐
70 to 79	SECOND-CLASS GUNNER ☐
69 or less	UNQUALIFIED ☐

13. VERIFIED BY

a. PRINTED NAME
1LT John B. Only

b. SIGNATURE
1LT John B. Only

c. GRADE
O-2

d. DATE (YYYYMMDD)
20171510

e. UNIT
HHC 2nd BN (MECH) 66th INF

DA FORM 5964, MAR 2017   PREVIOUS EDITIONS ARE OBSOLETE.

Figure 8-1. Sample of completed DA Form 5964

Chapter 8

8-8. Each unit armed with a mortar weapon system gives examinations semiannually. Other units may conduct examinations or allow their eligible members to take the qualification tests at nearby stations.

8-9. The commander authorized to issue special orders determines the date of the examination. The area selected should be on flat terrain consisting of soil that allows for aiming posts to be easily positioned at 50 and 100 meters from the station position.

8-10. The following personnel are eligible to take the examination:
- Commissioned officers, NCOs, and enlisted Soldiers assigned to a mortar unit.
- Commissioned officers, NCOs, and enlisted Soldiers whose duties require them to maintain proficiency in the use of mortars, as determined by battalion and higher commanders.

8-11. A candidate's earlier qualification ends when he is administered a record course with the mortar and is classified according to the latest examination score, as follows:
- Expert gunner: 90 to 100.
- First-class gunner: 80 to 89.
- Second-class gunner: 70 to 79.
- Unqualified: 69 or less.

GENERAL RULES

8-12. Conditions should be the same for all candidates during the test. The examining board ensures that information obtained by a candidate during testing is not passed to another candidate, and that candidates do not receive sight settings or laying of mortars left by a previous candidate.

8-13. The best available unit equipment should be used in the examination. Sight settings are considered correct when any part of the index coincides with any part of the line of graduation of the required setting.

8-14. The left side of the aiming post is used for alignment. The elevation and cross-level bubbles are considered centered when the bubbles are resting entirely within the outer etched lines on the vials.

8-15. The candidate is permitted to traverse the mortar to the middle point of traverse before each trial at laying the mortar, except at Station No. 5.

8-16. In any test that calls for mounting or emplacing the mortar, either by the candidate or the board, the surface emplacement is used. Digging is not allowed, and the rear of the baseplate assembly is not staked.

8-17. In time trials, the candidate does not receive credit for the trial if he performs any part of it after announcing, "UP."

8-18. The candidate selects assistants from within the squad to participate in the test. When squad members are unavailable for testing, the candidate may select assistants from outside the squad but from within the organization. The board makes sure that no unauthorized assistance is given the candidate during the examination.

8-19. A candidate is given three trials—one for practice and two for record. If the first trial is taken for record, then the second trial is taken for record even if candidate fails it. The credit score is the total of the two record trials. When the candidate fails in any trial through the fault of an examiner, defective sight, mortar, mount, or other instrument used, that trial is void and the candidate is given another trial as soon as possible. If the candidate's actions cause the mortar to function unsatisfactorily during testing, no credit is received for that portion of the test.

8-20. When a mechanical failure occurs and a mortar fails to maintain the lay after the candidate announces, "UP," a board member twists or pushes the mortar (taking up the play without manipulation) until the cross-level bubble is within the two outer etched lines. The board member then looks through the sight and if the vertical line is within two mils of the correct sight picture, the candidate is given credit for that trial, as long as other conditions are met.

8-21. The squad must repeat all commands. Commands should vary between trials, using even and odd numbers, and right and left deflections.

GROUND-MOUNTED MORTAR EXAMINATION

8-22. The ground-mounted mortar examination tests the gunner's ability to perform basic mortar gunnery tasks with the ground-mounted mortar system. It consists of the following tests with maximum credit scores, as shown:

- Mounting the mortar   20 points
- Small deflection change   20 points
- Referring the sight and realigning aiming posts (refer–realign)   20 points
- Large deflection and elevation changes   20 points
- Reciprocal lay   20 points

ORGANIZATION AND PROCEDURE

8-23. The recommended equipment needed for the five stations includes five mortars, five sights, one aiming circle, eight aiming posts, and five stopwatches.

8-24. The organization prescribed in table 8-1 is recommended for the conduct of the gunner's examination. Variations are authorized, depending on local conditions and the number of Soldiers being tested.

8-25. The candidate carries the scorecard (DA Form 5964) from station to station. The evaluator at each station fills in the time, trial scores, and credit score, and initials the appropriate spaces.

Chapter 8

Table 8-1. Conduct of ground-mounted gunner's examination

Station	Phase	Equipment	
		For Candidate	**For Examining Officer**
1	Mounting the mortar.	1 mortar 1 sight 1 baseplate stake	1 stopwatch
2	Small deflection change.	1 mortar 1 sight 2 aiming posts	1 stopwatch
3	Large deflection and elevation change.	1 mortar 1 sight 2 aiming posts	1 stopwatch
4	Refer - Realign.	1 mortar 1 sight 2 aiming posts	1 stopwatch
5	Reciprocal lay.	1 mortar 1 sight 2 aiming posts	1 stopwatch 1 aiming circle

MOUNTING THE MORTAR

8-26. The candidate is tested at Station No. 1 on his ability to perform the gunner's duties in mounting the mortar. The equipment needed is prescribed in table 8-1, page 8-4.

8-27. The candidate is directed to mount the mortar with his authorized assistants. The conditions of the test are described in the paragraphs below.

8-28. The candidate arranges the equipment as outlined in figures 8-2 through 8-4 on this page through page 8-7. The emplacement is marked before the examination.

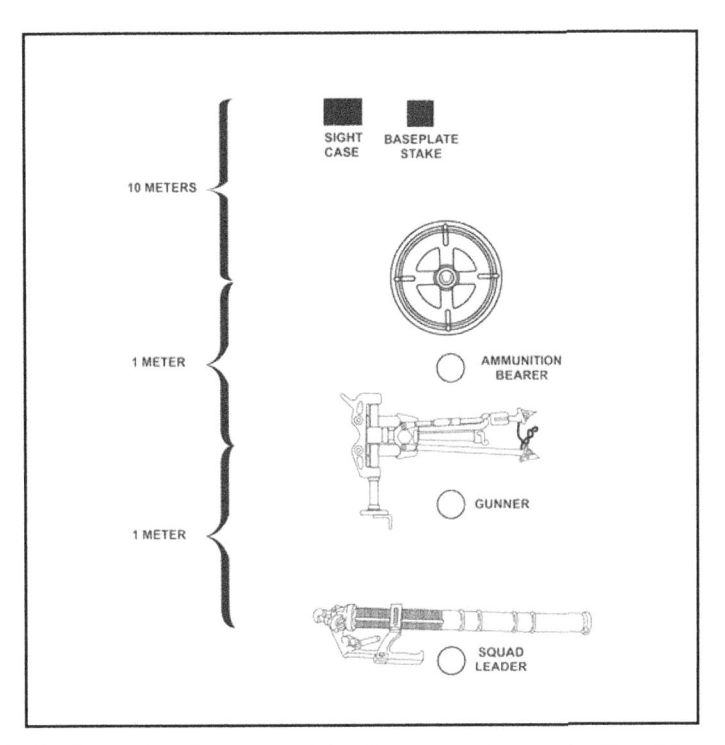

Figure 8-2. Equipment and personnel for the gunner's examination (60-mm mortar)

Chapter 8

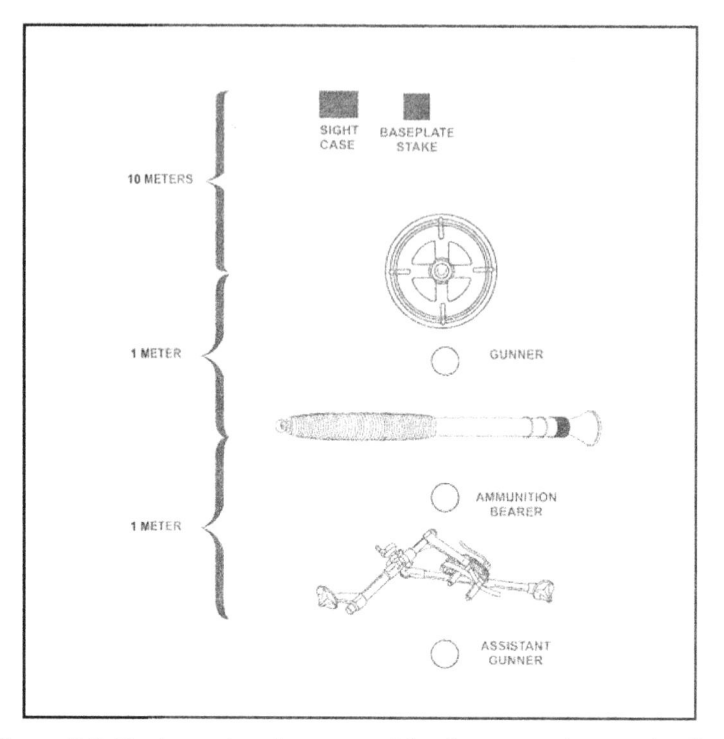

Figure 8-3. Equipment and personnel for the gunner's examination (M252/M252A1 81-mm mortar)

Gunner's Examination

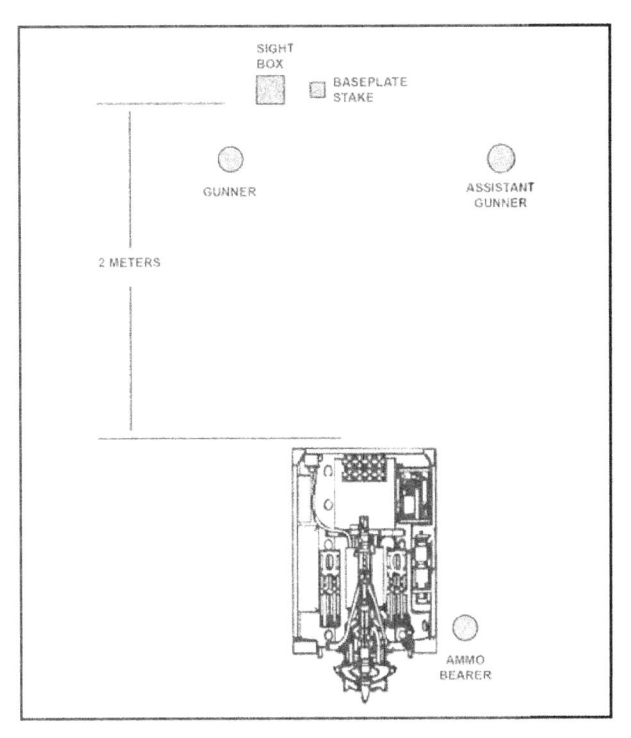

Figure 8-4. Equipment and personnel for the gunner's examination
(120-mm mortar with mortar stowage kit)

8-29. The mortar sight is seated in its case with 3800 mils set on the deflection scale and 1100 mils set on the elevation scale, and the sight box is closed and latched. For the 120-mm mortar, the traverse extension is locked and centered.

Procedure

8-30. The candidate is given three trials. If he chooses to use the first practice as a record, then the second is also for record. If he chooses to use the first trial as practice, the second and third trials are records. The credit score for the test is the total of the two record trials. The trials include the following procedure:

- The candidate and assistants take their positions. The candidate is instructed to mount the mortar at 3200 mils deflection and 1100 mils elevation.

- The evaluator points to the exact spot where the mortar is to be mounted and indicates the initial direction of fire by pointing in that direction. The evaluator gives the command "ACTION," at which time the candidate begins mounting the mortar. After mounting the mortar, the candidate should have 3200 mils deflection 1100 mils elevation.

- After the sight is mounted, ONLY THE GUNNER can manipulate the deflection and elevation settings on the M64A1/M67 sight unit, the traversing handwheel, the cross-leveling handwheel, and the crank.

Chapter 8

- The assistants may manipulate the sight mount knob/cross-level handwheel and elevation hand crank. They may center the connection for the mortar locking pin assembly, but they MUST NOT manipulate the sight for deflection or elevation settings.
- Time begins at the command of ACTION and time ends on the gunner announcing, "UP."

Scoring

8-31. Scoring procedures are as follows:
- The candidate receives no credit when the—
    - Time exceeds one minute, 15 seconds for the 120-mm, or 90 seconds for all other mortars.
    - Sight is not set correctly for deflection and elevation.
    - Cross-level and elevation bubbles are not centered.
    - Mortar locking pin or the clevis lock pin is not fully locked.
    - Connection for the mortar locking pin assembly (buffer carrier, 60-mm mortar) or the traversing slide assembly is off center more than two turns.
    - Assistant manipulates the sight for a deflection or elevation setting, after the sight is mounted.
    - Baseplate is not positioned correctly in relation to the baseplate stake.
    - Selector switch on the barrel is not on D for drop-fire (60-mm mortar only).
    - Collar assembly is not positioned on the lower saddle (60-mm mortar only).
    - Firing pin recess is not facing upwards on the barrel (M252/M252A1 81-mm mortar only).
    - Traverse is more than four turns (120-mm only).
    - Barrel clamp is not locked (120-mm only).
    - Cross-level lock is not tight (120-mm only).
    - Leg-locking handwheel is not wrist-tight (M252/M252A1 81-mm mortar only).
    - Coarse cross-level nut is not wrist-tight (60-mm mortar only).
    - Collar locking knob (60-mm mortar, hand-tight) or collar locking latch (81-mm mortar) is not secured to the barrel
    - Bipod legs are not fully extended and the spread cable or chain is not taut (60-mm and 120-mm mortars only).

- When the mortar is mounted correctly within the prescribed limits, credit is given as follows:

| Time in Seconds || Point Credit for Each Trial |
120-mm	60/81-mm	
51 or less	65 or less	10
52 to 57	66 to 70	9
58 to 63	71 to 75	8
64 to 69	76 to 80	7
70 to 75	81 to 85	6
-	86 to 90	5
76 or over	91 or over	0

SMALL DEFLECTION CHANGE

8-32. The candidate is tested at Station No. 2 on his ability to perform the gunner's duties when he is given commands that require a change in deflection. Equipment is prescribed in table 8-1, on page 8-4.

Conditions

8-33. A mortar is mounted with the sight installed. The sight is laid on two aiming posts (placed out 50 to 100 meters from the mortar) on a referred deflection of 2800 mils and 1100 mils elevation. The mortar is center of traverse, and the vertical line of sight is on the left edge of both aiming posts. The conditions of the test are as follows:
- The candidate is allowed to check the deflection set on the sight before each trial.
- The candidate is allowed to start each trial with his hand on the deflection knob.
- The change in deflection does not involve movement of the bipod assembly but causes the candidate to traverse the mortar at least 20 mils, but not more than 60 mils.
- Traversing extension is locked and centered (120-mm).

Procedure

8-34. The candidate is given three trials. If he chooses to use the first practice as a record, then the second is also for record. If he chooses to use the first trial as practice, the second and third trials are records. The credit score for the test is the total of the two record trials. The candidates follow the below guidelines:
- The candidate is given one assistant. A different command is given for each trial. The evaluator records the time and checks the candidate's work after each command has been executed.
- The evaluator announces an initial command requiring a change in deflection of 20 to 60 mils and an elevation change of 35 to 90 mils. The candidate may proceed with the exercise as soon as the deflection element is announced. The evaluator announces the command in normal sequence and cadence.
- The assistant is allowed to manipulate the elevation handwheel and crank.
- Time is charged against the candidate from the announcement of the last digit of the elevation element until the candidate announces, "UP."

Chapter 8

Scoring

8-35. Scoring procedures are as follows:
- The candidate receives no credit when the—
  - Time is 36 seconds or greater.
  - Sight is not set correctly for deflection or elevation.
  - Elevation bubble is not centered between the outer red lines.
  - Cross-level bubble is not centered between the outer red lines.
  - Assistant manipulates the mortar or sight for elevation or deflection.
  - Vertical cross hair of the sight is more than two mils off the correct sight picture.
- When the mortar is laid correctly within the prescribed limits, credit is given as follows:

Time in Seconds (All Mortars)	Point Credit for Each Trial
20 or less	10
21 to 23	9
24 to 26	8
27 to 31	7
32 to 35	6
36 or over	0

REFERRAL OF SIGHTS AND REALIGNMENT OF AIMING POSTS

8-36. The candidate is tested at Station No. 3 on his ability to perform the gunner's duties in referring the sight and realigning the aiming posts. Equipment is prescribed in table 8-1, page 8-4.

Conditions

8-37. The mortar is mounted with the correct sight installed. The sight is laid on two aiming posts (placed out 50 and 100 meters from the mortar) on a referred deflection of 2800 mils and 1100 mils elevation. The conditions of the test are as follows:
- The mortar is within two turns of center of traverse (four turns for the 120-mm). The candidate receives an administrative command with a deflection of 2860 or 2740 mils. The mortar is then re-laid on the aiming posts using the traversing crank and cross level handwheel.
- The candidate checks the conditions before each trial and is allowed to start the test with his hand on the deflection knob of the sight.
- The change in deflection in the command must be less than 25 mils but greater than 5 mils. The elevation remains constant at 1100 mils.
- The candidate is allowed two assistants—one to place out aiming posts and one to move the bipod (mount). The assistant gunner may manipulate the bipod for elevation. The assistant may center-up the traversing bar for the shift after the aiming posts have been realigned. The assistant may not manipulate the sight in any way.
- Traverse extension is not used. It remains locked in the center position.

Procedure

8-38. The candidate is given three trials. If he chooses to use the first practice as a record, then the second is also for record. If he chooses to use the first trial as practice, the second and third trials are records. The credit score for the test is the total of the two record trials. The candidates follow the below guidelines:

- A different command is given for each trial. The evaluator records the time and checks the candidate's work after each command has been executed.
- When the candidate is ready, he is given a command (for example), "REFER, DEFLECTION TWO EIGHT EIGHT ZERO, REALIGN AIMING POSTS."
- The candidate repeats each element of the command, sets the sight with the data given in the command, and directs one assistant in realigning the aiming posts. Then, the candidate centers the traversing assembly, and with the help of the assistant gunner, moves the bridge or bipod (mount) assembly and lays again on the aiming posts. After laying the mortar on the realigned posts, he announces, "UP."

Note. This procedure ensures that, after a registration mission (using a parallel sheaf), the mortars have matching deflections.

- Time is taken from the announcement of refer and align aiming post to the candidate's announcing, "UP."
- The candidate's assistant may not leave the mortar position until hearing the word POSTS in the command, "REALIGN AIMING POSTS."

Scoring

8-39. Scoring procedures are as follows:
- No credit is given when the—
  - Time exceeds one minute, 21 seconds.
  - Traversing crank is turned before the aiming posts are realigned.
  - Sight is not set correctly for deflection or elevation.
  - Mortar is not cross-leveled or correctly laid for elevation.
  - Vertical line of the sight is more than one mil off the correct sight picture.
  - Traversing assembly slide is more than two turns (four turns for the 120-mm) to the left or right of the center position.
  - Assistant manipulates the sight or mortar for deflection.
  - Aiming posts are aligned improperly.

Note. If for any reason either of the aiming posts fall before the candidate announces, "UP," the trial will be terminated and readministered.

Chapter 8

- When the mortar is found to be correctly laid within the prescribed limits, credit is given as follows:

Time in Seconds (All Mortars)	Point Credit for Each Trial
65 or less	10
66 to 70	9
71 to 75	8
76 to 80	7
81 or over	0

LARGE DEFLECTION AND ELEVATION CHANGES

8-40. The candidate is tested at Station No. 4 on his ability to perform the gunner's duties when he is given commands requiring a large change in deflection and elevation. Equipment is prescribed in table 8-1, page 8-4.

Conditions

8-41. A mortar is mounted with the sight installed. The sight is laid on two aiming posts placed out 50 to 100 meters from the mortar on a referred deflection of 2800 mils and elevation of 1100 mils. The elevation change is greater than 100 mils but less than 200 mils. The mortar is within two turns of center of traverse (four turns for the 120-mm). The conditions for the test are as follows:

- The candidate is allowed to check the deflection and elevation setting before each trial and is allowed to start each trial with his hand on the deflection knob.
- The change in deflection involves movement of the bipod assembly and causes the candidate to shift the barrel not less than 200 mils but not more than 300 mils. The change in elevation causes him to elevate or depress the barrel from low range to high range, or vice versa.
- Traverse extension is locked and centered (for the 120-mm).

Procedure

8-42. The candidate is given three trials. If he chooses to use the first practice as a record, then the second is also for record. If he chooses to use the first trial as practice, the second and third trials are records. The credit score for the test is the total of the two record trials. The candidate follows these guidelines:

- The candidate is given two assistants—one assistant may visually align the mortar, while the other shifts the bipod. The candidate maintains the option to shift the bipod without assistance. The assistants neither manipulate the sight nor lay the mortar for deflection. A different command is given for each trial. The evaluator records the time and checks the candidate's mortar after each command has been executed.
- The evaluator announces a command for a change in deflection and elevation that requires the movement of the bipod assembly, and a change in the elevation range—for example: "NUMBER ONE, HE QUICK, ONE ROUND, DEFLECTION THREE ZERO FOUR FIVE, CHARGE, FOUR, ELEVATION, ONE ONE FOUR ZERO."
- The candidate repeats each element of the command. As soon as the deflection element is given, he places the data on the sight and lays again on the aiming point with a compensated

sight picture. As soon as the mortar is laid, he announces, "UP." The assistants must remain in their normal positions until the deflection element is given.
- Time is taken from the announcement of the last digit of the elevation element of the fire command until the candidate announces, "UP."

Scoring

8-43. Scoring procedures are as follows:
- The candidate receives no credit when the—
    - Time exceeds time exceeds 55 seconds for the 120-mm or 60 seconds for all other mortars.
    - Sight is not set correctly for deflection or elevation.
    - Mortar is not correctly laid for elevation.
    - Mortar is not cross-leveled.
    - Vertical line is more than two mils off the compensated.
    - Traversing assembly slide is more than two turns (four turns for the 120-mm) to the left or right of the center position.
    - Assistants make unauthorized movements or manipulations.
    - Collar assembly is not positioned on the correct saddle for the announced elevation (60-mm).
    - Traverse extension is not locked and centered (for the 120-mm).
- When the mortar is laid correctly within the prescribed limits, credit is given as follows:

Time in Seconds		Point Credit for Each Trial
120-mm	**60/81-mm**	
35 or less	35 or less	10
36 to 40	36 to 40	9
41 to 45	41 to 45	8
46 to 50	46 to 50	7
51 to 55	51 to 55	6
-	56 to 60	5
56 or over	61 or over	0

RECIPROCAL LAY

8-44. The candidate is tested at Station No. 5 on his ability to perform the gunner's duties in laying a mortar for direction. Equipment is prescribed in table 8-1 on page 8-4.

Station Setup

8-45. The evaluator sets up the aiming circle about 25 meters to the left front of the station, levels the instrument, and orients the aiming circle so that the 0-3200 line is in the general direction the mortar is mounted. A direction stake is placed out about 25 meters in front of the mortar position.

Conditions

8-46. The candidate is given one assistant to shift the bipod assembly, but the candidate maintains the option to shift the bipod unassisted. The assistant does not manipulate the sight. The assistant may manipulate the bipod elevation and level the elevation bubble. The conditions of the test are as follows:

- All mortars are mounted at 3200 mils deflection and 1100 mils elevation. The mortar is laid on a direction stake on the initial mounting azimuth with the traversing mechanism centered.
- The mounting azimuth on which the candidate is ordered to lay the mortar is not less than 150 mils or more than 200 mils away from the initial mounting azimuth.
- The evaluator sets up the aiming circle about 25 meters to the left front of the mortar, with the instrument leveled and the 0-3200 line already on the mounting azimuth on which the mortar is to be laid.
- The candidate is allowed to start the test with his hand on the deflection knob. The assistants must remain in their normal positions until the evaluator gives the first deflection element.
- Traverse extension is locked and centered (120-mm).

Procedure

8-47. The candidate is given three trials. If he chooses to use the first practice as a record, then the second is also for record. If he chooses to use the first trial as practice, the second and third trials are records. The credit score for the test is the total of the two record trials. The candidate follows the below guidelines:

- The evaluator operates the aiming circle during this test, lays the vertical line on the mortar sight, and commands, "AIMING POINT THIS INSTRUMENT."
- The candidate refers his sight to the aiming point and replies, "AIMING POINT IDENTIFIED."
- The evaluator then announces the deflection—for example, "NUMBER ONE, DEFLECTION TWO THREE ONE FIVE."
- The candidate repeats the announced deflection, sets it on his sight, and lays the mortar on the center of the aiming circle lens. He then announces, "NUMBER ONE READY FOR RECHECK." The evaluator announces the new deflection immediately so that there is no delay.
- The operation is completed when the candidate announces, "NUMBER ONE, ZERO (or one) MIL(s), MORTAR LAID."
- Time is taken from the last digit of elevation first announced by the evaluator until the candidate announces, "NUMBER ONE, ZERO (or one) MIL(s), MORTAR LAID."

Scoring

8-48. Scoring procedures are as follows:

- The candidate receives no credit when the—
  - Time exceeds one minute, 51 seconds.
  - Sight is not set correctly for deflection or elevation.
  - Elevation bubble is not centered.
  - Cross-level bubble is not centered.
  - Vertical line of the sight is not centered on the aiming circle lens.
  - The mortar sight and the aiming circle deflection difference exceed one mil.

Gunner's Examination

- Assistant performs unauthorized manipulations or movements.
- Traversing mechanism is more than two turns (four turns for the 120-mm) from center of traverse. Traverse extension is not locked in the center position.

• When the mortar is laid correctly within the prescribed limits, credit is given as follows:

Time In Seconds (All Mortars)	Point Credit for Each Trial
50 or less	10
51 to 62	9
63 to 74	8
75 to 86	7
87 to 98	6
99 to 110	5
111 or over	0

M121 CARRIER-MOUNTED OR THE RMS6-L STRYKER-MOUNTED MORTAR EXAMINATION

8-49. This examination tests the gunner's ability to perform basic mortar gunnery tasks with the carrier-mounted 120-mm mortar system.

Note. All mounted platforms will use this examination in the conventional mode using the M-67 sight unit to demonstrate proficiency.

SUBJECTS AND CREDITS

8-50. The examination consists of the following tests, with maximum credit scores as shown:

Placing the mortar into a firing position from the traveling position.	20 points
Small deflection change.	20 points
Referring the sight and realigning the aiming posts.	20 points
Large deflection and elevation changes.	20 points
Reciprocal lay.	20 points

8-51. The minimum equipment needed for the five stations includes five mortars, five M1064A3-series carriers, five sights, one aiming circle, eight aiming posts, and five stopwatches.

8-52. The organization prescribed in table 8-2, page 8-16, is recommended for the conduct of the gunner's examination. Variations are authorized, depending on local conditions and the number of Soldiers being tested.

8-53. The candidate carries the scorecard from station to station. The evaluator at each station fills in the time, trial scores, and credit score, and initials the appropriate spaces.

Table 8-2. Conduct of carrier-mounted gunner's examination

Station	Phase	Equipment For Candidate	Equipment For Examining Officer
1	Placement of the mortar into firing position from the traveling position.	1 mortar carrier 1 mortar 1 sight	1 stopwatch
2	Small deflection change.	1 mortar carrier 1 mortar 1 sight 2 aiming posts	1 stopwatch
3	Referring of the sight and realignment of the aiming posts.	1 mortar carrier 1 mortar 1 sight 2 aiming posts	1 stopwatch
4	Large deflection and elevation changes.	1 mortar carrier 1 mortar 1 sight 2 aiming posts	1 stopwatch
5	Reciprocal lay.	1 mortar carrier 1 mortar 1 sight 2 aiming posts	1 stopwatch 1 aiming circle

PLACEMENT INTO A FIRING POSITION FROM TRAVELING 120-MM MORTAR

8-54. The candidate is tested at Station No. 1 on his ability to perform quickly and accurately the gunner's duties in placing the mortar into the firing position from the traveling position. Equipment is prescribed in table 8-2.

Conditions

8-55. The mortar is secured in the traveling position by the mortar tie-down strap. The conditions of the test are as follows:

- The sight is in its case, and the case is in its stowage position.
- The candidate selects an assistant gunner.
- The BAD is removed and stored properly for the 120-mm mortar system.
- The mortar hatch covers are closed and locked. The ramp may be in the up or down position.
- The gunner and assistant gunner are seated in their traveling positions.
- The evaluator ensures that the candidate understands the requirement of the test and instructs him to report, "I AM READY" before each trial.

Procedure

8-56. The candidate is given three trials. If he chooses to use the first practice as a record, then the second is also for record. If he chooses to use the first trial as practice, the second and third trials are records. The credit score for the test is the total of the two record trials. The candidate follows the below guidelines:

Note. The traverse extension is not used during the gunner's examination. It remains locked in the center position.

- The evaluator is positioned inside or outside the carrier to best observe the action of the candidate. The evaluator's position should not interfere with the action of the candidate.
- The trial is complete when the candidate announces, "UP."

Scoring

8-57. Scoring procedures are as follows:
- The candidate receives no credit when the—
    - Time exceeds one minute, 15 seconds.
    - Sight is not set at 3200 mils deflection and 1100 mils elevation.
    - Elevation and cross-level bubbles are not centered (within outer red marks).
    - Turntable and traversing assembly slide are not centered. The traverse extension must also be centered and locked.
    - The traversing lock handle is not locked.
    - The white line on the barrel is not aligned with the white line on the buffer housing assembly.
    - The mortar carrier rear hatch covers are not securely latched.
    - The safety mechanism is not set on FIRE. (F is showing.)
    - The cross-level locking knob is not hand-tight.
    - The buffer housing assembly is not positioned against the lower collar stop.
    - The BAD knob is not hand-tight.
    - The assistant manipulates the sight and mortar for deflection.
- When the mortar is found to be in the correct firing position within the prescribed limits, credit is given as follows:

Time in Seconds	Point Credit for Each Trial
50 or less	10
51 to 57	9
58 to 63	8
64 to 69	7
0 to 75	6
76 or over	0

SMALL DEFLECTION CHANGE

8-58. The candidate is tested at Station No. 2 on his ability to perform the gunner's duties when he is given commands that require a change in deflection. Equipment is prescribed in table 8-2 on page 8-16.

Conditions

8-59. The mortar is prepared for action with sight installed. The conditions of the test are as follows:
- The sight is laid on two aiming posts placed out 50 and 100 meters from the mortar on a referred deflection of 2800 mils and 1100 mils elevation. The turntable is centered, and the traversing mechanism is within four turns of center of traverse, and the traverse extension is centered and locked. The vertical line of the sight is on the left edge of both aiming posts.
- The change in deflection causes the candidate to traverse the mortar 20 to 60 mils for deflection and 30 to 90 mils for elevation.
- The candidate is allowed to begin the test with his hand on the deflection knob.

Procedure

8-60. The candidate is given three trials. If he chooses to use the first practice as a record, then the second is also for record. If he chooses to use the first trial as practice, the second and third trials are records. The credit score for the test is the total of the two record trials. The candidate follows the below guidelines:
- The evaluator announces an initial command requiring a change in deflection.
- The candidate repeats each element of the command, sets the sight with the data given, and traverses and cross-levels the mortar until obtaining the correct sight picture.
- Time is charged against the candidate from the announcement of the last digit of the elevation element until the candidate announces, "UP."

Scoring

8-61. Scoring procedures are as follows:
- The candidate receives no credit when—
  - The time exceeds 35 seconds.
  - The deflection or elevation is not indexed correctly.
  - The elevation and cross-level bubbles are not centered within the outer lines.
  - The vertical cross hair of the sight is not within one mil of the left edge of the aiming post.
  - The traverse extension is centered and locked in position.

- When the mortar is laid correctly within the prescribed limits, credit is given as follows:

Time in Seconds	Point Credit for Each Trial
20 or less	10
21 to 23	9
24 to 26	8
27 to 31	7
32 to 35	6
36 or over	0

REFERRAL OF SIGHTS AND REALIGNMENT OF AIMING POSTS

8-62. The candidate is tested at Station No. 3 on his ability to perform the gunner's duties in referring the sight and realigning the aiming posts. Equipment is prescribed in table 8-2, page 8-16.

Conditions

8-63. The sight is laid on two aiming posts (placed out 50 and 100 meters from the mortar) on a referred deflection of 2800 mils and 1100 mils elevation. The ramp is down with ammunition bearer in or outside the vehicle. The conditions of the test are as follows:
- The mortar is within four turns of center of traverse. The candidate receives an administrative command with a deflection of 2860 or 2740 mils. The mortar then is relaid on the aiming posts using the traversing crank.
- The candidate checks the conditions before each trial and is allowed to start the test with his hand on the deflection knob of the sight.
- The change in deflection in the command must be less than 25 mils but greater than 5 mils. The elevation remains constant at 1100 mils.
- The candidate selects two assistants—one assistant realigns the aiming posts and the other helps in moving the turntable and cross-leveling. The assistants do not manipulate the sight or mortar for deflection. The assistant may manipulate the elevation mechanism and level the elevation bubble.
- The traversing extension is not used. It remains locked in the center position.

Procedure

8-64. The candidate is given three trials. If he chooses to use the first practice as a record, then the second is also for record. If he chooses to use the first trial as practice, the second and third trials are records. The credit score for the test is the total of the two record trials. The candidate follows the below guidelines:
- A different command is given for each trial. The evaluator records the time and checks the candidate's work after each command has been executed.
- When the candidate is ready, he is given a command—for example, "REFER, DEFLECTION TWO EIGHT EIGHT ZERO, REALIGN AIMING POSTS."
- The candidate repeats each element of the command, sets the sight with the data given in the command, and directs one assistant in realigning the aiming posts. Upon completion of these

Chapter 8

actions, the candidate centers the traversing assembly. The assistant may center up the traversing assembly at the candidates discretion, and with the help of the other assistant, moves the turntable and lays again on the aiming posts. After laying the mortar on the realigned aiming posts, he announces, "UP."

- Time is taken from the announcement of "REFER, DEFLECTION TWO EIGHT EIGHT ZERO, REALIGN AIMING POSTS," until the candidate announces, "UP."
- The candidate's assistants are not permitted to leave the carrier until the command "REALIGN AIMING POSTS" is given.

Scoring

8-65. Scoring procedures are as follows:
- The candidate receives no credit when the—
  - Time exceeds one minute, 20 seconds.
  - Traversing assembly slide is turned before the aiming posts are realigned.
  - Traverse extension and turntable are not locked in the center position.
  - Sight is set incorrectly for deflection or elevation.
  - Elevation and deflection bubbles are not centered.
  - Sight picture is not correct.
  - Traversing assembly slide is more than four turns to the left or right of the center position.
  - Assistant manipulates the sight or mortar for elevation or deflection.
- When the mortar is laid correctly within the prescribed limits, credit is given as follows:

Time in Seconds	Point Credit for Each Trial
60 or less	10
61 to 65	9
66 to 70	8
71 to 75	7
76 to 80	6
81 or over	0

Note. If for any reason either of the aiming posts fall before the candidate announces, "UP," the trial will be terminated and readministered.

LARGE DEFLECTION AND ELEVATION CHANGES

8-66. The candidate is tested at Station No. 4 on his ability to perform the gunner's duties when given commands requiring a large change in deflection and elevation.

8-67. The evaluator selects a deflection change that is at least 200 but not more than 300 mils off the referred deflection of 2800 mils and 1100 mils elevation. The conditions of the test are as follows:

Gunner's Examination

- The change in deflection involves movement of the turntable. The change in elevation causes the candidate to elevate or depress the barrel from low range to high range, or vice versa. The change in elevation is not less than 100 mils but not more than 200 mils.
- The candidate selects two assistants.
- Traversing extension and turntable are locked at the center position.
- The candidate is allowed to check the deflection and elevation settings before each trial.
- The candidate is allowed to begin the test with his hand on the deflection knob.
- The assistant gunner may manipulate the elevation and level the elevation bubble.

Procedure

8-68. The candidate is given three trials. If he chooses to use the first practice as a record, then the second is also for record. If he chooses to use the first trial as practice, the second and third trials are records. The credit score for the test is the total of the two record trials. The candidate selects two assistants—one assistant may visually align the mortar, while the other elevates or depresses the standard assembly, and assists in moving the turntable. The assistant does not manipulate the sight or lay the mortar for deflection. The assistant may manipulate the elevation mechanism and level the elevation bubble. The candidate follows the below guidelines:

- The evaluator announces a command that requires a change in deflection involving movement of the turntable and an elevation change involving movement of the elevating mechanism cam.
- The candidate is allowed to start the test with his hand on the deflection knob, repeats each element of the fire command, and sets the sight with the data given in the command.
- As soon as the deflection element is announced, the candidate can immediately place the data on the sight. The assistants must remain in their normal positions until the elevation element is given.
- The evaluator times the candidate from the announcement of the last digit of the elevation command to the candidate announcing, "UP."
- A different deflection and elevation are given in the second trial.

Scoring

8-69. Scoring procedures are as follows:
- The candidate receives no credit when the—
  - Time exceeds 65 seconds.
  - Sight is not indexed correctly for deflection or elevation.
  - Elevation and cross-level bubbles are not centered.
  - Vertical line of the sight is more than two mils off the correct compensated sight picture.
  - Traversing mechanism is more than four turns off center of traverse.
  - Turntable is not in the locked position.
  - Assistants make any unauthorized manipulation of the mortar or sight unit for elevation or deflection.
  - Traversing extension is not locked in the center position.

Chapter 8

- When the mortar is laid correctly within the prescribed limits, credit is given as follows:

Time in Seconds	Point Credit for Each Trial
45 or less	10
46 to 50	9
51 to 55	8
56 to 60	7
61 to 65	6
66 or over	0

RECIPROCAL LAY (AIMING CIRCLE)

8-70. The candidate is tested at Station No. 5 on his ability to quickly and accurately perform the gunner's duties in reciprocal lay of the mortar. Equipment is prescribed in table 8-2, page 8-16.

Conditions

8-71. The mortar is prepared for action and laid on an initial azimuth by the evaluator and his assistants. The conditions of the test are as follows:
- The sight is set at 3200 mils deflection and 1100 mils elevation.
- The evaluator sets up the aiming circle about 75 meters from the carrier where it is visible to the gunner.
- The evaluator orients the aiming circle on an azimuth of not less than 150 mils but not more than 200 mils away from the initial azimuth.
- The candidate is allowed to begin the test with his hand on the deflection knob with the carrier engine running.
- A relay man is positioned halfway between the aiming circle and carrier to relay commands.
- The traversing mechanism is centered and the traversing extension is locked in the center position.

Procedure

8-72. The candidate is given three trials. If he chooses to use the first practice as a record, then the second is also for record. If he chooses to use the first trial as practice, the second and third trials are records. The credit score for the test is the total of the two record trials. The candidate follows the below guidelines:
- The evaluator operates the aiming circle during the test.
- Once the candidate identifies the aiming point, the evaluator announces the deflection.
- Time is started from the last digit of the first deflection announced by the evaluator.
- When the candidate announces, "READY FOR RECHECK," the evaluator immediately announces the new deflection.
- The assistant may manipulate the elevation mechanism and level the elevation bubble.
- The trial is complete when the gunner announces, "ZERO MILS (or ONE MIL), MORTAR LAID."

Scoring

8-73. Scoring procedures are as follows:
- The candidate receives no credit when the—
  - Time exceeds one minute, 35 seconds.
  - Difference between the deflection setting on the sight and the last deflection reading from the aiming circle is more than one mil.
  - Elevation and cross-level bubbles are not centered.
  - Vertical reticle line of the sight is not centered on the lens of the aiming circle.
  - Traversing extension is not locked in the center position.
  - The mortar sight and the aiming circle are not sighted on each other with a difference of more than one mil between deflection readings.
  - Turntable is not centered and locked.
- When the mortar is laid correctly, credit is given as follows:

Time in Seconds	Point Credit for Each Trial
55 or less	10
56 to 67	9
68 to 79	8
80 to 90	7
91 to 95	6
96 or over	0

RECIPROCAL LAY (COMPASS)

8-74. In this test (120-mm mortar only) the compass is substituted for the aiming circle.

8-75. The mortar is prepared for action and laid on an initial azimuth by the evaluator and his assistants. The conditions of the test are as follows:
- The turntable is centered with the sight set at 3200 mils deflection and 1100 mils elevation.
- The evaluator places the M2 compass on a stake about 75 meters from the mortar carrier and measures the azimuth to the mortar sight. The candidate then selects a mounting azimuth from the azimuth measured to the mortar sight.
- The candidate selects an assistant gunner and driver.
- The evaluator ensures that the candidate understands the requirements of the test and instructs him to report, "I AM READY," before each trial.

Procedure

8-76. The candidate is given three trials. If he chooses to use the first practice as a record, then the second is also for record. If he chooses to use the first trial as practice, the second and third trials are records. The credit score for the test is the total of the two record trials. The candidate follows the below guidelines:
- The evaluator operates the compass during the test.
- When the candidate identifies the aiming point, the evaluator announces the deflection.

Chapter 8

- When the gunner is laid back on the aiming point, he announces, "UP," and the evaluator commands, "REFER, DEFLECTION TWO EIGHT ZERO ZERO, PLACE OUT AIMING POSTS."
- The ammunition bearer moves out as soon as the initial deflection has been announced by the evaluator and places out the aiming posts as directed by the gunner.
- The assistant may manipulate the elevation mechanism and level the elevation bubble.
- The trial is complete when the gunner announces, "UP," after the aiming posts are in position.

Scoring

8-77. The scoring procedures are as follows:
- The candidate receives no credit when the—
    - Time exceeds two minutes, five seconds.
    - Deflection placed on the sight is incorrect.
    - Elevation and cross-level bubbles of the sight are not centered.
    - Turntable is not centered.
    - Aiming posts are not properly aligned.
    - Mortar is not within four turns of center of traverse.
- When the mortar is laid correctly, credit is given as follows:

Time in Seconds	Point Credit for Each Trial
70 or less	10
71 to 81	9
82 to 92	8
93 to 103	7
104 to 114	6
115 to 125	5
126 or over	0

RECIPROCAL LAY (SIGHT-TO-SIGHT)

8-78. In this test, the sight-to-sight method is used to reciprocally lay the mortar. The mortar is prepared for action on an azimuth by the evaluator and his assistants by taking the following actions:
- The sight is set at 3200 mils deflection and 1100 mils elevation.
- The evaluator sets up the base mortar about 35 meters from the test mortars where it is visible to the gunner.
- The evaluator orients the base mortar on an azimuth of not less than 150 mils or more than 200 mils away from the initial azimuth.
- The candidate selects an assistant gunner (optional for the 60-mm mortar).
- The candidate is allowed to begin the test with his hand on the deflection micrometer knob.
- The evaluator ensures that the candidate understands the requirements of the test, and instructs him to report, "I AM READY," before each trial.

Procedure

8-79. The candidate is given three trials. If he chooses to use the first practice as a record, then the second is also for record. If he chooses to use the first trial as practice, the second and third trials are records. The credit score for the test is the total of the two record trials. The candidate follows the below guidelines:

- The evaluator is positioned at the base mortar and commands, "AIMING POINT THIS INSTRUMENT."

- The gunner refers his sight to the aiming point and replies, "AIMING POINT IDENTIFIED."

- The evaluator reads the deflection from the sight of the base mortar, determines the back azimuth of that deflection by adding/subtracting 3200 mils, and announces the deflection—for example, the deflection on the base mortar is 1200 mils. The evaluator adds 3200 mils to this deflection (1200 + 3200 = 4400 mils) and announces, "NUMBER ONE, DEFLECTION FOUR FOUR ZERO ZERO."

- The candidate repeats the announced deflection, sets it on the sight, and, with the help of the assistant gunner, lays the mortar on the center of the base mortar sight lens, and announces, "NUMBER ONE READY FOR RECHECK." The evaluator announces the new deflection as soon as possible so that there is no delay.

- The assistant may manipulate the elevation mechanism and level the elevation bubble.

- The operation is completed when the candidate announces, "NUMBER ONE, ZERO (or ONE) MIL(S), MORTAR LAID."

Scoring

8-80. The scoring procedures are as follows:
- The candidate receives no credit when the—
  - Time exceeds one minute, 55 seconds.
  - Deflection placed on the sight is incorrect.
  - Elevation and cross-level bubbles of the sight are not centered.
  - Mortar is not within two turns of center of traverse.
  - The sight and the base mortar sight are not sighted on each other, with a difference of not more than one mil between deflection readings.
- When the mortar is laid correctly, credit is given as follows:

Time in Seconds	Point Credit for Each Trial
55 or less	10
56 to 67	9
68 to 79	8
80 to 91	7
92 to 103	6
104 to 115	5
116 and over	0

This page intentionally left blank.

Appendix A
Ammunition

Mortar ammunition component consists of a fuze, a cartridge, propellant charges and the fin assembly. A complete round of mortar ammunition contains all of the components needed to get the round out of the tube and to burst at the desired place and time. This appendix discusses the proper care and handling, field storage of ammunition and types of ammunition.

AMMUNITION COMPONENTS

A-1. U.S. mortars use a variety of **fuzes**:
- Point-detonation (PD), impact (IMP), or superquick (SQ) fuzes detonate the cartridge on impact with the ground.
- Near-surface burst (NSB) fuzes explode on or near the ground (0 to 1 meter).
- PRX fuzes explode above the ground. The 60/81/120 PRX detonates one to four meters above the ground.
- Delay (DLY) fuzes explode 0.05 seconds after impact with any object.
- Time fuzes explode after a preselected time has elapsed from the round being fired (illumination).
- Multioption fuzes combine two or more of the other modes into one fuze.

A-2. **Cartridges** can contain HE, ILLUM, or smoke, and are used for in the following situations:
- HE is used against personnel and light materiel targets.
- ILLUM (both conventional (visible) and infrared) are used in night missions requiring illumination for assistance in observation.
- Smoke, with red or white phosphorus (WP) filler, is used as a screening, signaling, casualty-producing, or incendiary agent.

A-3. **Propelling charges** provide the explosive force to propel the cartridge through the air. Propelling charges consist of a base charge (Charge 0) and from four to 10 incremental charges. The number of incremental charges used is determined by the range and time of flight.

A-4. The **fin assembly** consists of two components, the boom or shaft and the fin blades. (See figure A-1, page A-2.) The fin blades are at the base of the boom; it is important for the blades to be straight and have the appropriate thickness specified for the round. The fins stabilize the round in flight and cause it to strike fuze-end first. The fin shaft contains the propelling charge consisting of an ignition cartridge and removable propellant increments. The ignition cartridge (with primer) is fitted into the base of the tail fin shaft. The removable increments are fitted onto or around the shaft, depending on their type. The blades of some mortar rounds are canted to induce spin, which improving improves stability and dispersion.

Appendix A

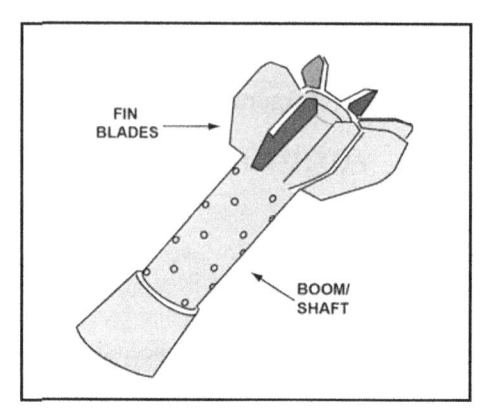

Figure A-1. Fin assembly

A-5. Fins around the tail of each cartridge stabilize it in flight and cause it to strike fuze-end first. The propelling charge consists of an ignition cartridge and removable propellant increments. The ignition cartridge (with primer) is fitted into the base of the fin shaft. The removable increments are fitted onto or around the shaft, depending on their type. When the cartridge is dropped fin-end first down the cannon, the ignition cartridge strikes the firing pin and detonates, causing a flash that passes through the radial holes in the shaft. This ignites the propelling increments, which produce rapidly expanding gases that force the cartridge from the barrel. The obturator band ensures equal muzzle velocities in hot or cold cannons by keeping gases inside the cannon until the cartridge has fired. When fired, the cartridge carries the ignition cartridge with it, leaving the mortar ready for the next cartridge.

AMMUNITION CARE AND HANDLING

A-6. The key to proper ammunition functioning is protection. Rounds prepared but not fired should be returned to their containers. Safety is always a matter of concern for all personnel and requires special attention where ammunition is concerned. Supervision is critical—improper care and handling can cause serious accidents, as well as inaccurate fire. Some of the principles of proper ammunition handling are—

- Never tumble, drag, throw, or drop individual cartridges or boxes of cartridges.
- Do not allow smoking, open flames, or other fire hazards around ammunition storage areas.
- Inspect each cartridge before it is loaded for firing. Dirty ammunition can damage the weapon or affect the accuracy of the round.
- Keep the ammunition dry, cool, out of the sun, and out of the elements, if possible.

CAUTION
Never make unauthorized alterations to components or mix propellants of one lot with another.

Note. For care and handling of specific mortar rounds, refer to the corresponding chapters in this publication.

Ammunition

> **DANGER**
> Never mix propelling charges.

A-7. Each projectile must be inspected to ensure that there is no leakage of the contents and that the projectile is correctly assembled.

BURNING UNUSED PROPELLING CHARGES

A-8. Mortar propelling charges are highly flammable and must be handled with extreme care to prevent exposure to heat, flame, or any spark-producing source, such as the hot residue from burning increments or propelling charges that float downward after a cartridge leaves the cannon. Like other types of ammunition, propelling charges must be kept cool and dry. Storing these items inside metal ammunition boxes until needed is an effective way to prevent premature combustion.

A-9. Unused charges must not be saved, but should be removed to a storage area until they can be burned or otherwise disposed of according to local range or installation regulations or SOPs.

A-10. Burning propellant charges may create a large flash, intense heat, and smoke. In a tactical environment, the platoon leader must ensure that burning increments do not compromise the security of the mortar position. The burning of propellant charges in a dummy position, if established, can aid in the deception effort. The safety officer, in a range environment, supervises the disposal of unused propellant charges.

FUZES

> **WARNING**
> Premature detonation may occur if a fuze is not properly seated.

A-11. To prevent accidental functioning of the point-detonating elements of M524-series fuzes, the fuzes must not be dropped, rolled, or struck under any circumstances. Any MT fuze that is modified after it is set must be reset to SAFE, and the safety wires (if applicable) must be replaced before the fuze is repacked in the original carton.

A-12. All primers must be inspected for signs of corrosion before use. If a seal has been broken, it is likely that the primer has been affected by moisture and it should be turned in.

AMMUNITION LOT NUMBER AND SEGREGATION OF AMMUNITION LOTS

A-13. Each ammunition lot is assigned an ammunition lot number, which is marked on each cartridge and packing container. It is used for records such as reports on condition, malfunctions, and accidents.

Appendix A

A-14. The markings on an ammunition container indicate its contents. Additional information is included on an ammunition data card inside each container. Each round is stenciled with the ammunition lot number, type of round, type of filler, and caliber.

A-15. Different lots of propellants burn at different rates and give slightly different effects in the target area. Therefore, the ammunition must be segregated by lot. In the field storage area, on vehicles, or in a dump, ammunition lots should be segregated and clearly marked.

AMMUNITION COLOR CODES

A-16. Mortar ammunition is painted and marked with a color code for quick, accurate identification. All mortar cartridges are painted to prevent rust and to identify their type.

A-17. A chart identifies rounds using NATO and U.S. color codes. (See table A-1.)

Table A-1. Mortar ammunition color codes

Type of Round	NATO Color Code			U.S. Color Code		
	Round	Markings	Band	Round	Markings	Band
High-explosive (HE) (Causes troop casualties and damage to light material)	Olive Drab	Yellow	NA (not applicable)	Olive Drab	Yellow	NA
White Phosphorous (WP) (To screen or signal, acts as an incendiary)	Light Green	Red	Yellow	Light Green	Red	Yellow
Red Phosphorous (RP) (To screen or signal, acts as an incendiary)	Light Green	Black	Brown	Light Green	Black	Brown
Illumination (ILLUM) (To illuminate, signal, and mark)	White	Black	NA	White	Black	Orange (infrared [IR] only)
Training Practice (For training and practice)	Blue	White	Brown or None	Blue	White	Brown

FIELD STORAGE OF AMMUNITION

A-18. Most ammunition components can be stored at temperatures as low as -50 degrees Fahrenheit for no longer than three days and as high as 145 degrees Fahrenheit for no more than four hours. Temperatures vary based on ammunition type.

A-19. Regardless of the method of storage, ammunition in the storage area faces certain hazards. The greatest hazards are weather; enemy fire; chemical, biological, radiological, nuclear

Ammunition

contamination; improper handling; and accidental fires. Adhere to the following guidelines to properly store ammunition:

- Stack ammunition by type and lot number. (See figure A-2.)

Note. 60-mm and 81-mm white phosphorus ammunition must be stacked fuze-end up so that the melted filler settles at the bottom, reducing the chance of voids forming off the long axis of the cartridge. Voids off the long axis of the cartridge cause the cartridge to fly erratically. 120-mm white phosphorus ammunition must not be stacked fuze-end down to prevent potential erratic flight of the ammunition when fired. Vertical, fuze end up, and horizontal stowage of 120-mm white phosphorous (WP) ammunition is permitted.

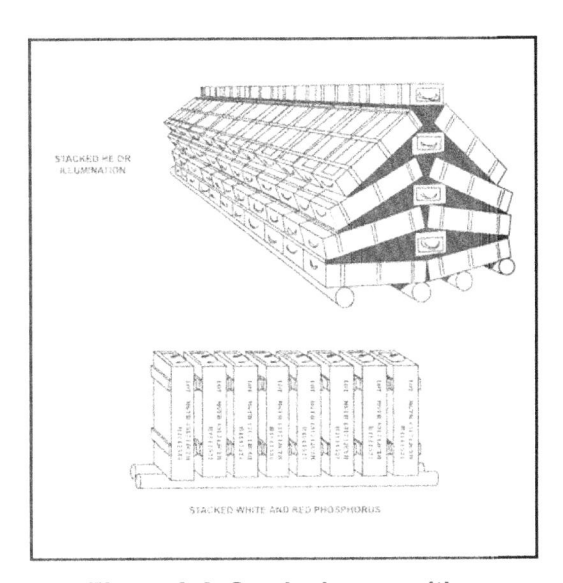

Figure A-2. Stacked ammunition

- If ammunition is stored on the ground, use at least six inches of strong dunnage under each stack.
- Keep the ammunition dry and out of direct sunlight by storing it in a vehicle or covering it with a tarpaulin. Be sure to provide adequate ventilation around the ammunition and between the covering material and the ammunition.
- Protect ammunition as much as possible from enemy indirect fires. If sandbags are used for protection, keep the walls at least six inches from the stacks and the roof at least 18 inches from the stacks to ensure proper ventilation.

A-20. With some PRX fuzes, the number of malfunctions can increase if fired in temperatures below 0 degrees Fahrenheit or above 120 degrees Fahrenheit. Powder temperature affects the muzzle velocity of a projectile and is of frequent concern to the FDC.

A-21. The four types of ammunition for the 60-mm mortar are high explosive (HE), white phosphorous (WP), illumination (ILLUM), and training. All 60-mm mortar cartridges, except training cartridges, have three major components—a fuze, a body, and a tail fin with a propulsion system assembly.

Appendix A

60-MM AMMUNITION

A-22. The ammunition used with the M224/M224A1 mortar can be classified by its use. Refer to the following tables (or TM 9-1010-233-10) for details on each cartridge used with the M224/M224A1.

Note. Mortar firing tables are available through the U.S. Army Research, Development and Engineering Command (RDECOM), Armament Research, Development and Engineering Center (ARDEC) who controls the publications management application (PMA) of Firing Tables and Ballistics: https://picac2as2.pica.army.mil/login/pma.htm

Firing Table 60-P-1 details all ammunition used with the M224/M224A1 mortar.

High-Explosive Ammunition

A-23. HE ammunition is used against enemy personnel and light materiel targets. Table A-2 on this page through A-8 details the HE ammunition that can be used when firing the M224/M224A1 mortar.

Table A-2. M224/M224A1 60-mm mortar high-explosive rounds

Cartridge/ Type	Maximum Range (Meters)	Fuze	Characteristics and Limitations
M720 HE cartridge	3490	M734 multioption fuze	- Temperature limits. Firing Lower limit -50°F Upper limit +145°F Storage: Lower limit -80°F (for period not more than 3 days). Upper limit +160°F (for period not more than 4 hr/day). - Do not fire the M720 cartridge with charge greater than 1 in the hand held mode.

Table A-2. M224/M224A1 60-mm mortar HE rounds (continued)

Cartridge/ Type	Maximum Range (Meters)	Fuze	Characteristics and Limitations
M720A1 HE cartridge	3490	M734A1 multioption fuze	- Temperature limits. Firing Lower limit -50°F Upper limit +145°F Storage: Lower limit -60°F Upper limit +160°F - Do not fire this cartridge above charge one in the handheld mode.
M720A2 HE cartridge	3490	M734A1 multioption fuze	- Temperature limits. Firing Lower limit -50°F Upper limit +145°F Storage: Lower limit -60°F Upper limit +145°F - Do not fire this cartridge above charge one in the handheld mode.
M888 HE cartridge	3490	M935 PD fuze	- Temperature limits. Firing Lower limit -50°F Upper limit +145°F Storage: Lower limit -80°F (for period not more than three days) Upper limit +160°F (for period not more than four hr/day) - This cartridge is identical to the M720, except that it has the M935 PD fuze. - Do not fire this cartridge above charge one in the handheld mode. *Note:* Cartridge must be fired a minimum distance of 300 meters during training.

Appendix A

Table A-2. M224/M224A1 60-mm mortar HE rounds (continued)

Cartridge/ Type	Maximum Range (Meters)	Fuze	Characteristics and Limitations
M768 HE cartridge	3490	M783 PD/ DLY fuze	- Temperature limits. Firing Lower limit -50°F Upper limit +145°F Storage: Lower limit -60°F Upper limit +160°F - Do not fire this cartridge above charge one in the handheld mode.
M768A1 HE cartridge	3490	M783 PD/ DLY fuze	- Temperature limits. Firing Lower limit -50°F Upper limit +145°F Storage: Lower limit -60°F Upper limit +145°F - Do not fire this cartridge above charge one in the handheld mode.
M1061 HE cartridge	3450	M734A1 multioption fuze	- Temperature limits. Firing Lower limit -50°F Upper limit +145°F Storage: Lower limit -60°F Upper limit +160°F

Note: HE cartridges cannot be fired above charge 1 in handheld mode and must be fired a minimum distance of 300 meters during training.

Legend: °F – degrees Fahrenheit, DLY – delay, hr – hour, HE – high explosive, PD – point detonating, WP – white phosphorous

Illumination Ammunition

A-24. ILLUM ammunition is used during night missions requiring assistance in observation. Table A-3 details the ILLUM ammunition that can be used when firing the M224/M224A1 mortar.

Table A-3. M224/M224A1 60-mm mortar illumination rounds

Cartridge/Type	Maximum Range (Meters)	Fuze	Characteristics and Limitations
M721 ILLUM cartridge	3490	M776 mechanical time, superquick (MTSQ) fuze	- Candle provides a minimum of 325,000 candlepower illumination for at least 40 sec - Temperature limits. Firing Lower limit -50°F Upper limit +145°F Storage: Lower limit -50°F (for period not more than three days) Upper limit +160°F (for period not more than four hr/day) - Do not fire below charge one.
M767 ILLUM cartridge	3490	M776 MTSQ fuze	- Candle provides a minimum of 500 candlepower illumination for at least 40 seconds. - Temperature limits. Firing Lower limit -50°F Upper limit +145°F Storage: Lower limit -50°F (for period not more than three days) Upper limit +160°F (for period not more than four hr/day) -This cartridge cannot be fired below charge two.
M83A3 ILLUM cartridge	931	M65A1-series time fuze	- Candle provides a minimum of 250,000 candlepower illumination for at least 30 sec. - Temperature limits. Firing Lower limit -40°F Upper limit +125°F Storage: Lower limit -80°F (for period not more than three days) Upper limit +160°F (for period not more than four hr/day) - Firing this cartridge below charge two will result in duds.
Legend: °F – degrees Fahrenheit, hr – hour, ILLUM – illumination, IR – infrared			

Appendix A

Smoke, White Phosporous Ammunition

A-25. Smoke, WP ammunition is used as a screening, signaling, or incendiary agent. Table A-4 details the smoke, WP ammunition that can be used when firing the M224/M224A1 mortar.

Table A-4. Smoke, white phosphorus ammunition for the M224/M224A1 60-mm mortar

Cartridge/Type	Maximum Range (Meters)	Fuze	Characteristics and Limitations
M722 smoke, WP cartridge	3490	M745 PD fuze	- Store and transport WP rounds at temperatures below 111.4°F (melting point of WP). It must be stored in a vertical position. - Temperature limits. Firing Lower limit -50°F Upper limit +145°F Storage: Lower limit -50°F (for period not more than three days) Upper limit +160°F (for period not more than 4 hr/day).
M722A1 smoke, WP cartridge	3490	M783 PD/ DLY fuze	- Store and transport WP rounds at temperatures below 111.4°F (melting point of WP). It must be stored in a vertical position. - Temperature limits. Firing Lower limit -50°F Upper limit +145°F Storage: Lower limit -50°F Upper limit +145°F

Table A-4. Smoke, white phosphorus ammunition for the M224/M224A1 60-mm mortar (continued)

Cartridge/Type	Maximum Range (Meters)	Fuze	Characteristics and Limitations
M302A1/A2 smoke, WP cartridge	1630	M527, M935, or M936 PD fuze	- Store and transport WP rounds at temperatures below 111.4°F (melting point of WP). It must be stored in a vertical position. - Temperature limits. Firing Lower limit -40°F Upper limit +125°F Storage: Lower limit -80°F (for period not more than three days) Upper limit +160°F (for period not more than 4 hr/day) - Short rounds may occur when fired in below 0 °F temperatures.
Legend: °F – degrees Fahrenheit, DLY – delay, hr – hour, PD – point detonating, WP – white phosphorous			

Full Range Practice Cartridge

A-26. The full range practice cartridge (FRPC) is used for training when service ammunition is not available or there are restrictions on the use of service ammunition for training. Table A-5 details the practice ammunition used when firing the M224/M224A1 mortar.

Table A-5. M224/M224A1 60-mm mortar practice rounds

Cartridge/Type	Maximum Range (Meters)	Fuze	Characteristics and Limitations
M769 FRPC	3490	M775 PD practice fuze	- The cartridge is similar in appearance to M720 and M720A1 HE cartridges. - Temperature limits. Firing Lower limit 0°F Upper limit +110°F Storage: Lower limit -50°F Upper limit +145°F
Legend: °F – degrees Fahrenheit, FRPC – full range practice cartridge, HE – high explosive, PD – point detonating			

Appendix A

81-MM AMMUNITION

A-27. Ammunition for the M252/M252A1 is typed according to its use. The tables outlined below the authorized cartridges for the M252/M252A1 81-mm mortar.

High-Explosive Ammunition

A-28. HE ammunition is used against enemy personnel and light materiel targets. Table A-6 details the HE ammunition that can be used when firing the M252/M252A1 mortar.

Table A-6. M252/M252A1 81-mm mortar HE rounds

Cartridge/ Type	Maximum Range (Meters)	Fuze	Characteristics and Limitations
M821, M821A1, M821A2 HE cartridges (C869)	5608 (M821) (M821A1) 5792 (M821A2)	M734 MO fuze (M821/A) M734A1 MO fuze (M821A2)	- Temperature limits. Firing Lower limit -50°F Upper limit +145°F Storage: Lower limit -60°F Upper limit +160°F
M889, M889A1, M889A2 HE cartridges (C869)	5536 (M889) 5844 (M889A1) 5792 (M889A2)	M935 PD fuze (M889) (M889A1) M783PD/ DLY (M889A2)	- Temperature limits. Firing Lower limit -50°F Upper limit +145°F Storage: Lower limit -60°F Upper limit +160°F
M374A2, M374A3 HE Cartridge (C256)	4500 (M374A2) 4800 (M374A3)	M524/M567 PD fuze M532 PROX (M374A2, M374A3)	-- Temperature limits. Firing Lower limit -50°F Upper limit +145°F Storage: Lower limit -60°F Upper limit +160°F
Legend: °F – degrees Fahrenheit, DLY – delay, hr – hour, HE – high explosive, PD – point detonating, WP – white phosphorous			

Ammunition

Illumination Ammunition

A-29. ILLUM ammunition is used during night missions requiring assistance in observation. Table A-7 details the ILLUM ammunition that can be used when firing the M252/M252A1 mortar.

Table A-7. M252/M252A1 81-mm mortar illumination rounds

Cartridge/Type	Maximum Range (Meters)	Fuze	Characteristics and Limitations
M853A1 ILLUM cartridge	5100	M772 MTSQ fuze	- Temperature limits. Firing Lower limit -50°F Upper limit +145°F Storage: Lower limit -60°F Upper limit +160°F
M816 IR ILLUM cartridge	5100	M772 MTSQ fuze	- Temperature limits. Firing Lower limit -50°F Upper limit +145°F Storage: Lower limit -60°F Upper limit +160°F
M301A1, M301A2 ILLUM cartridge	2150	M84 time fuze	- Temperature limits. Firing Lower limit -40°F Upper limit +125°F Storage: Lower limit -80°F (for period not more than three days) Upper limit +160°F (for period not more than four hr/day). - Firing with less than two propellant increment charges (charge two) is not authorized.

Appendix A

Table A-7. M252/M252A1 81-mm mortar illumination rounds (continued)

Cartridge/Type	Maximum Range (Meters)	Fuze	Characteristics and Limitations
M301A3 ILLUM cartridge	3150	M84A1 time fuze	- Temperature limits. Firing Lower limit -40°F Upper limit +125°F Storage: Lower limit -80°F (for period not more than three days) Upper limit +160°F (for period not more than four hr/day). - Firing with less than three propellant increment charges (charge three) is not authorized.
Legend: °F – degrees Fahrenheit, hr – hour, ILLUM – illumination, IR – infrared, MTSQ – mechanical time, superquick			

Ammunition

Smoke, White Phosporous and Red Phosphorous Ammunition

A-30. Smoke, WP, and red phosphorous ammunition is used as a screening, signaling, or incendiary agent. Table A-8 details the smoke WP and red phosphorous ammunition that can be used when firing the M252/M252A1 mortar.

Table A-8. M252/M252A1 81-mm mortar smoke, white phosphorus, and red phosphorous rounds

Cartridge/Type	Maximum Range (Meters)	Fuze	Characteristics and Limitations
M819 smoke red phosphorous cartridge	5000	M772 PD MTSQ fuze	- Temperature limits. Firing Lower limit -50°F Upper limit +145°F Storage: Lower limit -50°F Upper limit +160°F
M375A1 and M375A2 smoke, WP cartridge	4500	M524 PD fuze	- Temperature limits. Firing Lower limit -40°F Upper limit +125°F Storage: Lower limit -80°F (for period not more than three days) Upper limit +160°F (for period not more than four hour/day). - Store and transport WP rounds at temperatures below 111.4°F (melting point of WP). It must be stored in a vertical position.
M375A3 smoke WP cartridge	4800	M524A1 PD fuze	- Temperature limits. Firing Lower limit -40°F Upper limit +125°F Storage: Lower limit -80°F (for period not more than three days) Upper limit +160°F (for period not more than four hr/day). - Store and transport WP rounds at temperatures below 111.4°F (melting point of WP). It must be stored in a vertical position.
Legend: °F – degrees Fahrenheit, hr – hour, MTSQ – mechanical time, superquick, PD – point detonating, WP – white phosphorus			

Appendix A

Training Practice Ammunition

A-31. Training practice ammunition is used for training when service ammunition is not available or there are restrictions on the use of service ammunition for training. Table A-9 details the TP ammunition that can be used when firing the M252/M252A1 mortar.

Table A-9. M252/M252A1 81-mm mortar training and practice rounds

Cartridge/Type	Maximum Range (Meters)	Fuze	Characteristics and Limitations
M879 FRPC	5700	M751 PD practice fuze	- Temperature limits. Firing Lower limit 0°F Upper limit +110°F Storage: Lower limit -45°F Upper limit +145°F
Legend: °F – degrees Fahrenheit, FRPC – full range practice cartridge, PD – point detonating			

120-MM AMMUNITION

A-32. The ammunition used with M120, M121, and the RMS6-L 120-mm mortars can be classified by its use. The following cartridges can be fired from both M120 and M121 mortars, except for M91 ILLUM cartridges, which only can be fired from M120 and ground-mounted M121 mortars.

High-Explosive Ammunition

A-33. High-explosive ammunition is used against enemy personnel and light materiel targets. Table A-10 details the HE ammunition that can be used when firing M120 and M121 mortars.

Table A-10. M120 and M121 120-mm mortar high-explosive rounds

Cartridge/Type	Maximum Range (Meters)	Fuze	Characteristics and Limitations
M934/M934A1 HE cartridges	7200	M734 multioption (M934) fuze M734A1 multioption (M934A1) fuze	- Temperature limits. Firing Lower limit -50°F Upper limit +145°F Storage: Lower limit -60°F Upper limit +160°F
M933 HE cartridge	7200	M745 PD fuze	- Temperature limits. Firing Lower limit -50°F Upper limit +145°F Storage: Lower limit -60°F Upper limit +160°F
M57 HE cartridge	7200	M935 PD fuze	- Temperature limits. Firing Lower limit -28°F Upper limit +145°F Storage: Lower limit -50°F Upper limit +145°F
Note: All rounds have a minimum range of 200 meters at charge 0. (Charge may vary in firing table and whiz wheel.)			
Legend: °F – degrees Fahrenheit, HE – high explosive, PD – point detonating			

Appendix A

Illumination Ammunition

A-34. Illumination ammunition is used for battlefield illumination and signaling. Table A-11 details the ILLUM ammunition that can be used when firing M120 and M121 mortars.

Table A-11. M120- and M121 120-mm mortar illumination rounds

Cartridge/Type	Maximum Range (Meters)	Fuze	Characteristics and Limitations
M983 IR ILLUM cartridge	6675	M776 MTSQ fuze	- Temperature limits. Firing Lower limit -50°F Upper limit +145°F Storage: Lower limit -60°F Upper limit +160°F - Provides IR illumination for 50 seconds and is used to support troops with night vision devices.
M930 and M930E1 ILLUM cartridges	6800 (M930) 6975 (M930E1)	M776 MTSQ fuze	- Temperature limits. Firing Lower limit -50°F Upper limit +145°F Storage: Lower limit -60°F Upper limit +160°F - The filler in these cartridges provides an illumination of one million candlepower for 60 seconds.
M91 ILLUM cartridge	7100	M776 MTSQ fuze	- Temperature limits. Firing Lower limit -28°F Upper limit +145°F Storage: Lower limit -50°F Upper limit +145°F - The filler provides an illumination of one million candlepower for 60 seconds. - This cartridge can only be fired from M120 and ground-mounted M121 mortars. It weighs 27.01 pounds.
Legend: °F – degrees Fahrenheit, IR - infrared, ILLUM – illumination, MTSQ – mechanical time, superquick			

Ammunition

Smoke, White Phosporous Cartridges

A-35. Smoke cartridges with WP filler are used for screening and spotting. Table A-12 details the smoke WP ammunition that can be used when firing M120 and M121 mortars.

Table A-12. M120 and M121 120-mm mortar smoke, white phosphorus rounds

Cartridge/Type	Maximum Range (Meters)	Fuze	Characteristics and Limitations
M929/XM929 smoke WP cartridges	7200	M734A1 multioption (M929) fuze M745 PD (XM929) fuze	- Temperature limits. Firing Lower limit -50°F Upper limit +145°F Storage: Lower limit -60°F Upper limit +160°F - The filler provides an illumination of one million candlepower for 60 seconds. - Store and transport WP rounds at temperatures below 111.4°F (melting point of WP). It must be stored in a vertical position.
Note: All rounds have a minimum range of 200 meters at charge 0. (Charge may vary in firing table and whiz wheel.)			
Legend: °F – degrees Fahrenheit, PD – point detonating, WP – white phosphorous			

WARNING

At temperatures exceeding 111.4 degrees Fahrenheit (44.1 degrees centigrade) (melting point of WP), store and transport WP rounds in a vertical position (nose up) to prevent voids in the WP.

Appendix A

Training-Practice Ammunition

A-36.   TP cartridges are used when training squads and indirect fire teams (less expensive than service ammunition) or when the use of service ammunition is restricted. Table A-13 details the training ammunition that can be used when firing M120 and M121 mortars.

Table A-13. M120 and M121 120-mm mortar practice round

Cartridge/Type	Maximum Range (Meters)	Fuze	Characteristics and Limitations
M931 FRPC	7200	M781 PD fuze	- Temperature limits. Firing Lower limit 0°F Upper limit +110°F Storage: Lower limit -45°F Upper limit +145°F
Note. All rounds have a minimum range of 200 meters at charge 0. (Charge may vary in firing table and whiz wheel.)			
Legend: °F – degrees Fahrenheit, FRPC – full-range practice cartridge, PD – point detonating			

TYPES OF FUZES

A-37.   A fuze is a device to initiate a projectile at the time and under the circumstances required. Mortar fuzes are located at the top of the cartridge and have four possible actions: SQ acts on impact, DLY acts 0.05 seconds after impact, variable time acts a certain distance above the ground, and MT acts after a preset number of seconds have elapsed after being fired. Many modern fuzes have more than one action.

> **WARNING**
>
> Do not try to disassemble any fuze.

PROXIMITY FUZES

A-38.   A PRX fuze is an electronic device that detonates a projectile by means of radio waves sent out from a small radio set in the nose of the projectile. The M532 fuze is a radio Doppler fuze that has a PRX or SQ function. An internal clock mechanism provides nine seconds of safe air travel (610 to 2340 meters along trajectory for charge zero through nine, respectively). Once set to act as an IMP fuze, the mechanism cannot be reset for PRX. The fuze arms and functions normally when fired at any angle of elevation between 0800 and 1406 mils at charges one through nine. The fuze is not intended to function at charge zero.

A-39.   However, at temperatures above 32 degrees Fahrenheit and at angles greater than 1068 mils, the flight time is sufficient to permit arming. To convert the fuze from PRX to SQ, the top of the fuze must be rotated 120 degrees (one-third turn) in either direction. This action breaks an internal shear pin and internal wire, which disables the proximity function.

> **CAUTION**
>
> Do not attempt to reset the M532 variable time fuze. Fuzes set for point detonating cannot be returned to proximity mode.

A-40.   Proximity fuzes withstand normal handling without danger of detonation or damage when in their original packing containers, or when assembled to projectiles in their packing containers.

> **WARNING**
>
> The explosive elements in primers and fuzes are sensitive to shock and high temperatures. Do not drop, throw, tumble, or drag boxes containing ammunition.

Disposal Procedures

A-41.   PRX-fuzed short cartridges that are duds contain a complete explosive train and impact element. They should not be approached for five minutes or disturbed for at least 30 minutes after misfiring.

A-42.   After 30 minutes, the dud is still dangerous but can be approached and removed carefully or destroyed in place by qualified disposal personnel. If the situation allows for a longer waiting period, the dud is considered safe for handling after 40 hours.

Burst Height

A-43.   The principal factors affecting height of burst are the angle of approach to the target and the reflectivity of the target terrain. The air burst over average types of soil ranges from one to six meters, depending on the angle of approach.

A-44.   High angles of approach (near vertical) give the lowest burst heights. Light tree foliage and light vegetation affect the height of burst only slightly, but dense tree foliage and dense vegetation increase the height of burst. Target terrain such as ice and dry sand gives the lowest burst heights, whereas water and wet ground give the highest burst heights.

A-45.   Close approach to crests, trees, towers, large buildings, parked aircraft, mechanized equipment, and similar irregularities creates the need for above average bursting heights. When targets are beyond such irregularities, a clearance of at least 30 meters should be allowed to ensure maximum effect over the target area.

A-46.   The fuzes may be used for day or night operations. They function normally in light rain; however, heavy rain, sleet, or snow can increase the number of early bursts. At extreme

Appendix A

temperatures (below -40 degrees Fahrenheit and above 125 degrees Fahrenheit), it is not unusual to experience an increase in malfunctions proportionate to the severity of conditions.

MECHANICAL TIME FUZES

A-47. These fuzes use a clockwork mechanism to delay functioning for a specific time.

M734 Multioption Fuze

A-48. This **60-mm M734 multioption fuze** (see figure A-3) can be set to function as a proximity burst (one to four meters above the surface), near-surface burst (zero to one meter above the surface), impact/superquick burst (on the surface), or delay burst (0.05 seconds).

A-49. It is set by hand without using a tool. Once set, the M734 setting can be changed many times before firing without damaging the fuze.

Note. This fuze has no safety pins or wires.

A-50. If a cartridge set for PRX fails to burst at the proximity distance above the target, it automatically bursts at zero to one meter (zero to three feet) above the target. If a cartridge set for NSB fails to burst at the near-surface distance above the target, it automatically bursts on impact. If a cartridge set for impact fails to burst on impact, it automatically bursts 0.05 seconds after impact.

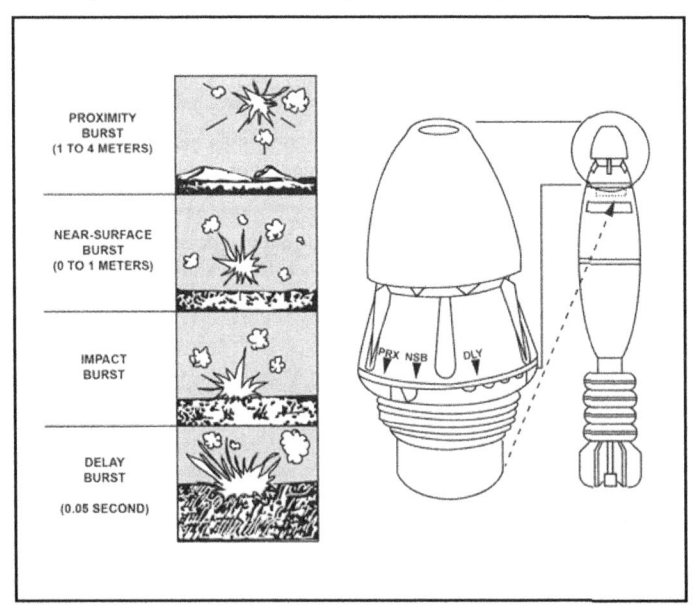

Figure A-3. M734 multioption fuze

Ammunition

A-51. The **81-mm M734 multioption fuze** has four function settings:
- PRX causes the cartridge to explode between one and three meters above the ground.
- NSB causes the cartridge to explode up to one meter above the ground.
- IMP causes the cartridge to explode on contact.
- DLY incorporates a 0.05-second delay in the fuze train before exploding the cartridge.

A-52. No tools are needed to set the fuze. The setting can be changed several times without damaging the fuze. It has no safety pins or wires, which reduces preparation time. If the fuze does not function as set, it automatically functions at the next lower setting.

A-53. The **120-mm M734 multioption fuze** (see figure A-4) can be set by hand and has the following characteristics:
- Functions: PRX, NSB, IMP, or DLY.
- Settings: PRX, NSB, IMP, or DLY.

A-54. The fuze can be set by rotating the fuze head (clockwise) until the correct marking (PRX, NSB, IMP, or DLY) is over the index line. There are selectable functions that the fuze can be set to—
- PRX. The fuze comes set to PRX. Burst height is three to 13 feet (one to four meters).
- NSB (nonjamming). Burst height is zero to three feet (zero to one meter).
- IMP (SQ).
- DLY (0.05 seconds).

A-55. Rotate the fuze head (counterclockwise) until the PRX marking is over the index line resets the fuze.

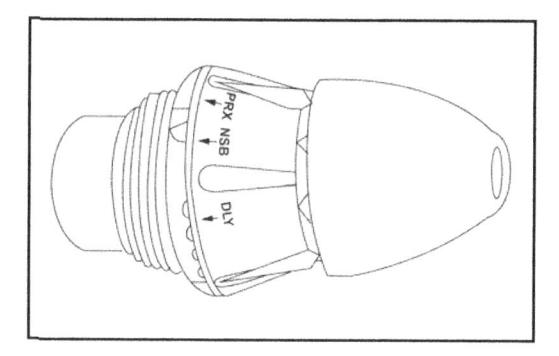

Figure A-4. M734 multioption fuze setting

M734A1 Multioption Fuze

A-56. The air-powered **60-mm M734A1 multioption fuze** has four selectable functions:
- PRX (60/81 PRX setting is with the 60-mm).
- 120 PRX (not used with the 60-mm).
- IMP.
- DLY.

A-57. In HE PRX mode, the height of burst (HOB) remains constant over all types of targets. The IMP mode causes the round to function on contact with the target and is the first backup

Appendix A

function for either PRX setting. In the DLY mode, the fuze functions about 30 to 200 milliseconds after target contact. The DLY mode is the backup for the IMP and PRX modes. The IMP and DLY modes have not changed from the current M734 multioption fuze.

A-58. Radio frequency jamming can affect the functioning of PRX fuzes. Radio frequency jamming initiates a gradual desensitizing of the fuze electronics to prevent premature fuze function. Once the fuze is out of jamming range, the fuze electronics recover and function in the PRX mode if the designed HOB has not been passed. To limit the time of fuze radio frequency radiation, the proximity turn-on is controlled by an apex sensor that does not allow initiation of the fuze proximity electronics until after round has passed the apex of the ballistic trajectory.

A-59. The safety requirements of military standard 1316C, require that the M734A1 uses ram air and setback to provide two independent environment sensors.

A-60. The air-powered **81-mm M734A1 multioption fuze** has four selectable functions—
- 60/81 PRX.
- 120 PRX (not used with the 81-mm).
- IMP.
- DLY.

A-61. In HE PRX mode, the HOB remains constant over all types of targets. The impact mode causes the round to function on contact with the target and is the first backup function for either PRX setting. In the DLY mode, the fuze functions about 30 to 200 milliseconds after target contact. The DLY mode is the backup for the IMP and PRX modes. The IMP and DLY modes have not changed from the current M734 multioption fuze.

A-62. Radio frequency jamming can affect the functioning of PRX fuzes. Radio frequency jamming initiates a gradual desensitizing of the fuze electronics to prevent premature fuze function. Once the fuze is out of jamming range, the fuze electronics recover and function in the PRX mode if the designed HOB has not been passed. To limit the time of fuze radio frequency radiation, the proximity turn-on is controlled by an apex sensor that does not allow initiation of the fuze proximity electronics until after the round has passed the apex of the ballistic trajectory.

A-63. The safety requirements of military standard 1316C require that the M734A1 uses ram air and setback to provide two independent environment sensors.

A-64. The air-powered **120-mm M734A1 multioption fuze** has four selectable functions—
- PRX 120.
- PRX 60/81.
- IMP.
- DLY.

A-65. In HE PRX mode, the HOB remains constant over all types of targets. The IMP mode causes the round to function on contact with the target and is the first backup function for either PRX setting. In the DLY mode, the fuze functions about 30 to 200 milliseconds after target contact. The DLY mode is the backup for the IMP and PRX modes. The IMP and DLY modes have not changed from the current M734 multioption fuze.

A-66. Radio frequency jamming can affect the functioning of PRX fuzes. Radio frequency jamming initiates a gradual desensitizing of the fuze electronics to prevent premature fuze function. Once the fuze is out of jamming range, the fuze electronics recover and function in the PRX mode if the designed HOB has not been passed. To limit the time of fuze radio frequency

Ammunition

radiation, the proximity turn-on is controlled by an apex sensor that does not allow initiation of the fuze proximity electronics until after the round has passed the apex of the ballistic trajectory.

A-67. The safety requirements of military standard 1316C require that the M734A1 uses ram air and setback to provide two independent environment sensors.

M935 Point-Detonating Fuze

A-68. The **60-mm M935 PD fuze** can be set to function as SQ or DLY.

A-69. It is preset for the SQ function, and the ammunition handler verifies that the selector slot is aligned with the SQ mark on the ogive. The ammunition handler selects DLY by turning the slot clockwise until it is aligned with the "D" marking on the ogive.

A-70. Its standard pull-wire safety is removed immediately before firing.

A-71. The **81-mm M935 PD fuze** has two function settings: IMP and a 0.05-second DLY. It is set using the bladed end of the M18 fuze wrench. The M935 is fitted with a standard pull wire and safety pin that are removed immediately before firing.

A-72. All PD fuzes are SQ and, therefore, detonate on impact. The **120-mm M935 PD fuze** (see figure A-5) has a safety wire and the following characteristics:
- Functions: Impact.
- Settings: SQ or 0.05-second delay action.

A-73. These fuzes are preset to function superquick on impact. Verify the setting before firing. The selector slot should be aligned with the SQ mark. To set the delay action, turn the selector slot in a clockwise direction until the slot is aligned with the DLY mark. Use a coin or a flat-tip screwdriver to change settings.

A-74. Align the selector slot with the SQ mark to reset the fuze.

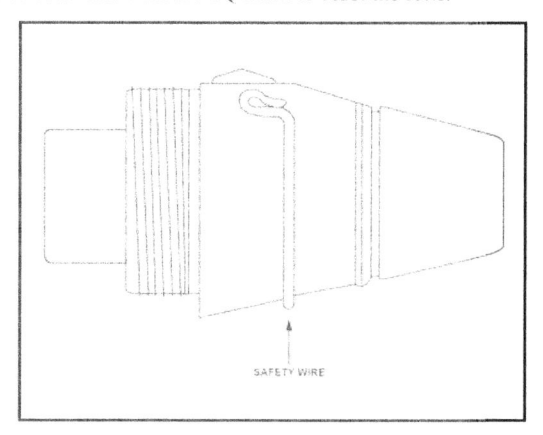

Figure A-5. M935 point-detonating fuze

FUZE, M524-SERIES

A-75. The **81-mm M524-series fuze** has two function settings: SQ and DLY. When set at DLY, the fuze train causes a 0.05-second delay before functioning. When set at SQ, the fuze

Appendix A

functions on point impact or graze contact. The fuze contains a delayed arming feature that ensures it remains unarmed and detonator safe for a minimum of 1.25 seconds and a maximum of 2.50 seconds of flight.

A-76. To prepare for firing, the slot is aligned in the striker with SQ or DLY using the M18 fuze wrench. The safety pull wire is removed just before inserting the cartridge into the mortar.

Note. If upon removal of the safety wire, a buzzing sound in the fuze is heard, the round should not be used. The round is still safe to handle and transport if the safety wire is reinserted.

WARNING

If the plunger safety pin (upper pin) cannot be reinserted, the fuze may be armed. An armed fuze must not be fired since it will be premature. It should be handled with extreme care and explosive ordnance (EOD) personnel notified immediately. If it is necessary to handle a round with a suspected armed fuze, personnel must hold the round vertically with the fuze striker assembly up.

M526-Series Fuze

A-77. The **81-mm M526-series fuze** has an SQ function only. It is fitted with a safety wire and pin that are removed immediately before firing.

M567 Fuze (81-mm) and M783 Point-Detonating and Delay Fuze (60-mm and 81-mm)

A-78. The M567 and M783 fuze act only on impact. The **81-mm M567 fuze** is a selective SQ or 0.05-second delay impact fuze. It comes preset to function as SQ, and the selector slot should align with the SQ mark on the ogive. To set for delay, the selector slot should be rotated clockwise until it is aligned with the "D" mark on the ogive. An M18 fuze wrench is used to change settings. The fuze has a safety wire that must be removed before firing.

A-79. The **M783 point-detonating and delay fuze** (60-mm and 81-mm) can be set to function as SQ or DLY. M783 fuze is a point-detonating/delay fuze. (See figure A-6.) If the M783 fuze is set on 120 PRX or 60/81 PRX, it only will function in impact mode.

Ammunition

> **WARNING**
>
> Cartridge fuzed with the M783 are known to detonate prematurely on very rare occasions. Detonations have occurred in the upward and downward portions of flight, but outside of the fuze's arming distance (100 meters).

A-80. Functions: Impact/delay. Settings: Dummy multioption, 120 PRX, 60/81 PRX, IMP (impact), or DLY (0.05 second-delay) action.

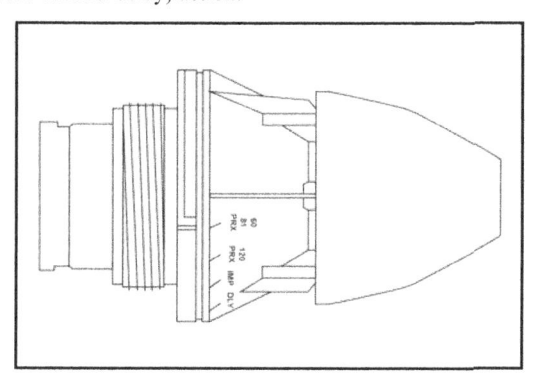

Figure A-6. M783 point-detonating and delay fuze

M745 Point-detonating Fuze

A-81. The **M745 point-detonating fuze** (see figure A-7) functions on impact with SQ action only. Disregard the markings (PRX, NSB, IMP, and DLY) on the fuze head. The M745 fuze has the following characteristics:
- Functions: Impact.
- Settings: None. (No setting or resetting is required.)

Appendix A

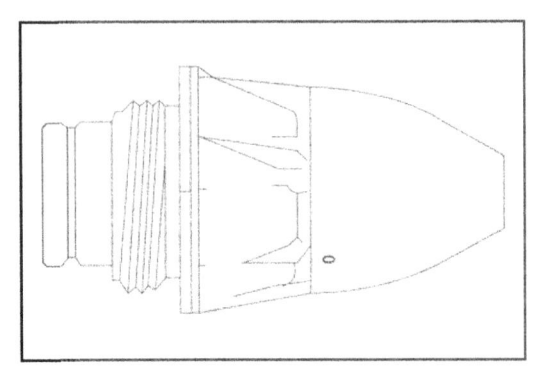

Figure A-7. M745 point-detonating fuze

M65-Series Time Fuze

A-82. The **M65** is a fixed-time fuze and has a safety wire.

M776 Mechanical Time Superquick Fuze

A-83. Using a fuze setter, the ammunition handler sets the time on the fuze, and once fired, the cartridge explodes after that amount of time has elapsed. If the mechanical time fails, the fuze acts on impact. The safety wire is removed just before firing.

M779 Point-detonating Fuze and M775 Practice Fuze

A-84. The **M779 point-detonating fuze** and **M775 practice fuze** are fuzes are used with M766 and M769 training cartridges only.

> **WARNINGS**
>
> **Do not** fire cartridges with M525 or M527 PD fuzes if **the fuze** makes a buzzing sound when the safety pins **are rem**oved. Check the fuze for the presence of the **bore-rid**ing pin after removing the safety pin and do **not fire** the cartridge if it is missing. Notify EOD.
>
> **Adequa**te fragmentation cover must be taken when **firing** cartridges for distances less than 300 meters
>
> **Soldiers** should consult the training manual on firing **temperature** limits.

Ammunition

> **CAUTION**
> Examine fuze for visible damage or looseness. Do not attempt to retighten a loose fuze or reinstall a detached fuze. Cartridges with damaged, loose, or detached fuzes must be turned in to the ammunition supply point (ASP) as unserviceable

M772 Fuze

A-85. The **M772 fuze** is a mechanical time, superquick (MTSQ) fuze that can be set from three to 55 seconds at half-second intervals. The settings are obtained from the range tables and are applied using a wrench (No. 9239539) or a 1 3/4-turn open-end wrench. The safety wire must be removed before firing.

M84 Fuze

A-86. The **M84 fuze** is a single-purpose, powder-train, MT fuze used with the 81-mm M301A1 and M301A2 illuminating cartridges. It has a time setting of up to 25 seconds. The fuze consists of a brass head, body assembly, and expelling charge. The fuze body is graduated from zero to 25 seconds in one-second intervals; five-second intervals are indicated by bosses. The zero-second boss is wider and differs in shape from the other body bosses; the safe setting position is indicated by the letter "S" on the fuze body.

A-87. The adjustment ring has six raised ribs for use in conjunction with M25 fuze setter and a setting indicator rib (marked SET) about half the height and width of the other six ribs. Safety before firing is provided by a safety wire, which must be removed just before firing.

M84A1 Fuze

A-88. The **M84A1 fuze** is a single-purpose, tungsten-ring, MT fuze used with the 81-mm M301A3 ILLUM cartridge. It has a time setting of up to 50 seconds. All other features are the same as the M84 fuze.

Fuze Wrench

A-89. The **fuze wrench, M18**, assembles the fuze to the cartridge, and the bladed tip on the end sets PD-type fuzes. The fuze setting wrench (NSN: 5120-00-203-4801) sets M772 MT and M768 time fuzes. It engages the 1 3/4-inch flats on the setting ring or the fuze head. The M25 fuze setter sets M84-series time fuzes. Notches in the setter engage ribs in the setting ring of the fuze.

M751 Practice Fuzes

A-90. The **M751 and the M775 practice fuzes** are used with the 81-mm mortar. There are two types of M751 fuzes:
- Type 1 resembles the M734 fuze.
- Type 2 resembles the M935 fuze.

Appendix A

A-91. The M751 is fitted with a smoke charge that operates on impact. The safety/packing clip should be removed when the cartridge is unpacked.

M776 Mechanical Time, Superquick Fuze

A-92. The **M776 MTSQ fuze** (see figure A-8, page A-30) has a mechanical arming and timing device, an IMP fuze, an expulsion charge, a safety wire or pin, and the following characteristics.
- Functions: air burst or impact.
- Settings: six to 52 seconds.

A-93. Use the fuze setter to rotate the head of the fuze to the left (counterclockwise) until the inverted triangle or index line is aligned with the correct line and number of seconds of the time scale. If the desired setting is passed, continue counterclockwise to reach the desired setting. Try to index the desired setting in as few rotations as possible.

A-94. To reset the fuze, rotate the head of the fuze (counterclockwise) until the safe line (S or inverted triangle of the time scale) is aligned with the index line of the fuze body. Replace the safety wire.

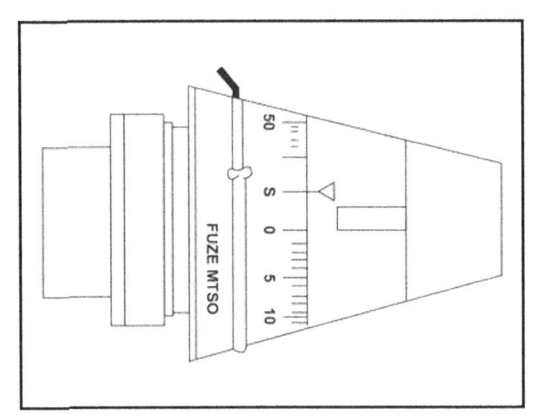

Figure A-8. M776 mechanical time

CARTRIDGE PREPARATION

A-95. Cartridges are prepared as close to the actual firing period as possible. The propellant train (except the training cartridge) consists of an ignition cartridge and propellant charges. The ignition cartridge has a percussion primer and is assembled to the end of the fin assembly. The propelling charge is contained in four horseshoe-shaped, felt-fiber containers or nine wax-treated, cotton cloth, bag increments. The propelling charges are assembled around the fin assembly shaft.

A-96. Cartridges are shipped with a complete propelling charge, an ignition cartridge, and a primer. Firing tables are used to determine the correct charge for firing. Remaining increments are repositioned toward the rear of the tail fin assembly when firing the cartridge with horseshoe-shaped increments at less than full charge.

Ammunition

Note.
> Charge zero—ignition cartridge only.
> Charge one—ignition cartridge and one increment.
> Charge two—ignition cartridge and two increments.
> Charge nine—ignition cartridge and nine increments.

A-97. Increments removed from cartridges before firing should be placed in a metal or wooden container located at least 25 meters away from the firing vehicle or position. Excess increments should not accumulate near the mortar positions but are removed to a designated place of burning and destroyed. Units should follow specific range SOPs to dispose of unused increments. For example—

- Select a place at least 100 meters from the mortar position, parked vehicles, and ammunition points.
- Clear all dead grass or brush within 30 meters around the burning place. Do not burn increments in piles. Spread them in a train one to two inches deep, four to six inches wide, and as long as necessary.
- From this train, extend a starting train that will burn against the wind of single increments laid end to end. End this starting train with not less than one meter of inert material (dry grass, leaves, or newspapers).
- Ignite the inert material.
- Do not leave unused increments unburned in combat operational areas. The enemy will use them.

UNPACKING

A-98. Standard cartridges are packed individually in fiber containers with eight cartridges in a metal container. Use proper tools to open ammunition containers. Unpack the cartridge and remove any packing material. Do not break the moisture-resistant seal of the shipping container until the ammunition is to be used. When a large number of cartridges (15 or more for each squad) are prepared before a combat mission, the cartridges may be removed from the shipping container and the propellant increments adjusted. Reinsert the fin assemblies into the container to protect the propelling charges.

- When opening an ammunition box, the ammunition bearer ensures the box is horizontal to the ground, not nose-end or fin-end up. After the bands are broken and the box opened, the rounds should be removed by allowing them to roll out along the lid of the box. After the rounds have been removed, always use two hands to prevent accidental dropping. Dropping may cause the propellant charges to ignite, causing bodily injuries.
- To help minimize the occurrence of short rounds and duds, unpackaged ammunition that has been dropped should not be fired. It should be destroyed in accordance with standard procedures.

Appendix A

> **WARNING**
>
> Incidents occurring from mishandling 300-series ammunition have resulted in minor burns to the hands and legs.

Note. The floating firing pin within the primer has about 1/16 of an inch to move around. This may cause the firing pin to ignite the charges if the cartridge is dropped on the fin end. (See figure A-9.)

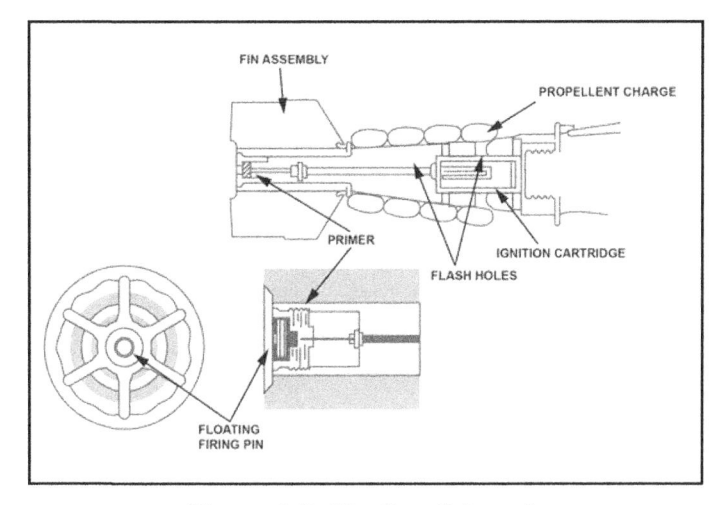

Figure A-9. Floating firing pin

A-99. Always handle complete cartridges with care. The explosive elements in primers and fuzes are sensitive to shock and high temperature.

A-100. The moisture-resistant seal of the container is broken when the ammunition is to be used. When a large number of cartridges are needed for a mission, they may be removed from the containers and prepared. Propelling charges are covered or protected from heat and moisture.

A-101. The ammunition is protected from mud, sand, dirt, and water. If it gets wet or dirty, it must be wiped off at once. The powder increments should not be exposed to direct sunlight. Keeping ammunition at the same temperature results in a more uniform firing.

A-102. The pull wire and safety wire are removed from the fuze just before firing. When cartridges have been prepared for firing, but are not used, all powder increments and safety wires are replaced. The cartridges are returned to their original containers. These cartridges are used first in subsequent firing so that once-opened stocks can be kept to a minimum.

A-103. The four types of ammunition for the 120-mm mortar are HE, smoke, ILLUM, and training. All mortar cartridges, except training cartridges, are packaged as a complete round and have three major components—a fuze, body, and tail fin with propulsion system assembly.

CARTRIDGE INSPECTION

A-104. Examine the cartridge fin assembly for any visible damage or looseness. Examine the fuze and propelling charges for any visible damage. Tighten loose fin assemblies by hand. Cartridges with damaged (bent) fin assemblies, fuzes, or propelling charges are turned in to the ASP as unserviceable.

Appendix A

> **WARNING**
>
> The M781 fuze on the M931 FRPC may be armed if the packing clip is missing or the red band on the striker is protruding from the nose cap. Any force applied to the nose of an armed fuze can result in ignition of the propelling charge. Do not attempt to fire the cartridge. Remove it to the dud pit without striking the nose of the fuze.

CARTRIDGE HANDLING

A-105. To properly handle cartridges, the squad uses the following procedures:
- Do not throw or drop live ammunition.
- Do not break the moisture resistant seal of the ammunition container until the cartridges are to be fired.
- Protect cartridges when removed from the ammunition container. Protect ammunition from rain and snow. Do not remove the plastic shell or insert assembly around the propelling charge until the propelling charge is to be adjusted. If protective bags were packed with the cartridge, cover the fin assembly and propelling charge to prevent moisture contamination. Stack cartridges on top of empty ammunition boxes or on four to six inches (10 to 15 centimeters) of dunnage. Cover cartridges with the plastic sheets provided.
- Do not expose cartridges to direct sunlight, extreme temperatures, flame, or other sources of heat.
- Shield cartridges from small-arms fire.
- Store WP-loaded cartridges at temperatures below 111.4 degrees Fahrenheit (44.1 degrees centigrade) to prevent melting of the WP filler. If this is not possible, WP-loaded cartridges must be stored fuze-end up. After the temperature returns below 111.4 degrees Fahrenheit (44.1 degrees centigrade) WP will resolidify with the void space in the nose end of the cartridge. (Failure to observe this precaution could result in rounds with erratic flight.
- Store WP-loaded ammunition separately from other types of ammunition.
- Notify EOD of leaking WP cartridges. Avoid contact with any cartridges that leak.
- Protect the primer of cartridges during handling.
- Do not handle duds other than when performing misfire procedures.

A-106. Ammunition is manufactured and packed to withstand all conditions normally encountered in the field. However, since explosives are affected by moisture and high temperature, they must be protected. Explosive elements in primers and fuzes are sensitive to strong shock and high temperature. Complete cartridges being fired must be handled with care. Adhere to the following guidelines for proper ammunition care and handling:
- Do not throw or drop cartridges.
- Protect ammunition from mud, dirt, sand, water, snow, and direct sunlight. Cartridges must be free of such foreign matter before firing. Ammunition that is wet or dirty should be wiped off at once.

Ammunition

> **WARNING**
>
> Do not walk on, tumble, drag, throw, roll, or drop ammunition. Ensure that ammunition is kept in the original container until ready for use. Do not combine WP and HE in storage. Maintain compatibility and quantity of ammunition, as outlined in Technical Bulletin (TB) 43-0250.
>
> Unpacked mortar cartridges are known to sustain significant damage when dropped. Do not fire unpacked ammunition which has been dropped, or packaged ammunition which has been dropped from a height greater than one meter.
>
> Defects incurred may not be detectable by visual inspection, and firing damaged ammunition could result in weapon or property damage, or serious personnel injury or death.
>
> All cartridges which have been dropped are to be segregated, tagged or marked for identification immediately, and turned in to the ASP as unserviceable. These cartridges will be returned to ASP as unserviceable. Packed ammunition dropped from a height of more than seven feet could be similarly damaged.

- Always store ammunition under cover. When this is not possible, raise it at least six inches (15 centimeters) off the ground and cover with a double thickness of tarpaulin. Dig trenches around the ammunition pile for drainage. All WP cartridges must be stored with the fuze-end up and follow the recommendations below in the appropriate situation:
    - In combat, store ammunition underground such as in bunkers.
    - In the field, use waterproof bags, ponchos, ground cloths, and dunnage to prevent deterioration of ammunition. Ensure that ammunition does not become water-soaked.
    - In arctic weather, store the ammunition in wooden boxes or crates. Place the boxes or crates on pallets and cover them with a double thickness of tarpaulin.
    - Cover fin assemblies and propelling charges with the fiber container or end cap. Stack cartridges on top of empty ammunition boxes and cover them with plastic sheets.
- Take the following actions to handle WP cartridges:
    - Store WP cartridges at temperatures below 111 degrees Fahrenheit to prevent the WP from melting. When temperatures are above 111 degrees Fahrenheit, store the WP with the fuze-end up. This prevents the WP filler from forming voids away from the cartridge's designed center of mass and causing an erratic flight of the cartridge.
    - Store WP cartridges away from other types of ammunition.
    - Do not handle duds. Follow local range procedures to dispose of them.

Appendix A

> **WARNING**
>
> Duds are cartridges that have been fired but have not exploded. Duds are dangerous and should not be handled by anyone other than an EOD team member.

A-107. Before-firing, checks include the following:
- Ammunition should be free of moisture, rust, and dirt.
- The fin and fuze assembly is checked for tightness and damage.
- Charges are kept dry.
- Extra increments are removed if the cartridge is to be fired with less than full charge.
- With the exception of a few unused increments (within the same ammunition lot number) as replacements for defective increments, excess powder should be removed from the mortar position.
- The primer cartridge is checked for damage or dampness.

PREPARING TO FIRE

A-108. When the ammunition bearer receives information in a fire command, he prepares the ammunition for firing. The number of cartridges, type of cartridge, fuze setting, and charge are all included in the fire command. To apply the data, the ammunition bearer selects the proper cartridge, sets the fuze, and adjusts (removes or replaces) the number of propelling charges on the quantity of cartridges called for in the fire command. Each cartridge is inspected for cleanliness and serviceability. The ammunition bearer also ensures that all packing material is removed and that there is no glue or other foreign substance adhering to the cartridge, especially around the obturator band. Any safety wires are pulled just before firing.

Ammunition

> **WARNING**
>
> For protection, a cartridge prepared but not fired should be returned to its container, increment end first. The pull wire on the M935 fuze on the M888 cartridge must be replaced before returning it to its container.
>
> Always check the mask and overhead clearance before firing.
>
> PD and PRX fuzes may function prematurely when fired through extremely heavy rainfall.
>
> Do not fire ammunition in temperatures above +146 degrees Fahrenheit (+63 degrees centigrade) or below -28 degrees Fahrenheit (-3 degrees centigrade).
>
> Short rounds and misfires can occur if an excessive amount of oil or water is in the cannon during firing.
>
> Before loading the cartridge, ensure the cannon and cartridge are free of sand, mud, moisture, snow, wax, or other foreign matter. Ensure that all packing materials (packing stops, supports, and plastic bags) are removed from the cartridge. Remove any glue or other foreign substances adhering to the cartridge, particularly at or near the obturator band. If the substance cannot be removed, the cartridge should not be fired.

Note. The ammunition bearer immediately adjusts the number of charges in a fire for effect (FFE) mission. In an adjust-fire mission, the ammunition bearer prepares the cartridge but delays adjusting the charges until the FFE is entered, because the charge may change during the subsequent adjustment of fire.

PROPELLING CHARGE ADJUSTMENTS

A-109. The ammunition bearer adjusts the number of charges immediately in an FFE mission. In an adjust-fire mission, he prepares the cartridge and delays adjusting the charges until the FFE is entered because the charge may change during the subsequent adjustments of fire. Cartridges are shipped with a complete propelling charge consisting of the ignition cartridge and four charges. The fire command includes the number of charges on the cartridge.

A-110. If the cartridge is fired at less than full charge, the ammunition handler removes the number of charges based on the fire command. The charges are as follows:

- Charge zero is the ignition charge only.
- Charge one has the ignition charge and one increment.

Appendix A

- Charge two has the ignition charge and two increments.
- Charge three has the ignition charge and three increments.
- Charge four has the ignition charge and four increments.

A-111. If the charge is less than charge four, slide the remaining increments towards the rear until they are positioned against the fins. Place the excess increments in an empty ammunition box for protection and close the lid during firing. Dispose of the excess increments according to local range regulations.

A-112. Communication among the mortar squad is imperative in order to mitigate the probability of human error when selecting the appropriate round type, number of charges, and fuze setting. Whenever possible, the squad leader physically inspects each round prior to firing. If the situation does not facilitate physical inspection, the squad leader must visually inspect the round and verbally confirm and acknowledge with the assistant gunner that the round or rounds are all correct. The gunner remains as the last control measure to verify that all data for the fire mission is without error (deflection and elevation, sight picture, and round). If any squad member suspects an unsafe condition during firing, it is that member's responsibility to call, "CHECK FIRE."

> **WARNINGS**
>
> Propelling charges are not interchangeable. Do not substitute one model for another. Do not mix lots.
>
> Never fire cartridges with a greater number of propelling charges than authorized for the ammunition and weapon.
>
> Reposition remaining propellant increments toward rear of fin assembly. Failure to comply may result in serious injury to personnel or damage to equipment.

UNFIRED CARTRIDGES

A-113. If a cartridge is not fired, replace the safety wire (if it was removed from the fuze) and reset the fuze. Reinstall the propellant increments so that the cartridge has a full charge. Repack the cartridge in its original packaging.

Notes. If the safety pins cannot be inserted fully into the fuze, notify EOD.

M68 and M91 cartridges should have seven propelling charges. The correct order is one brown, two blue, and four white. Use the original increments and do not mix propelling charge models or lots.

> **CAUTION**
>
> Do not attempt to reset the M532 variable time fuze. Fuzes set for PD cannot be returned to proximity mode.

Appendix B
Training Devices

The most efficient and direct method of teaching conduct of fire is by firing service ammunition under field conditions. However, ammunition training allowances and range limitations often restrict such training. When properly used, mortar training devices provide realistic training at reduced costs and at more accessible locations.

B-1. The M931 FRPC provides realistic training for all members of the indirect fire team at a reduced cost. The M931 FRPC is a nondud-producing, 120-mm training cartridge used to simulate the M934 HE cartridge for M120 and M121 120-mm mortars.

B-2. The weight, charges, fuzing, and ballistic characteristics of the cartridge are similar to those of the HE cartridge. The M781 fuze for the M931 produces a flash, a bang, and a smoke signature that provide audio and visual feedback to the mortar squad and FO. The M931 has the following general characteristics:
- Maximum range: 7200 meters.
- Minimum range: 200 meters.
- Maximum rate of fire: 16 rounds per minute (first minute).
- Sustained rate of fire: Four rounds per minute.

B-3. The cartridge allows the entire mortar team—FO, FDC, and guns—to practice the actions required for a fire mission. These actions include—
- Forward observer CFF and adjustment.
- FDC calculations and procedures.
- Mortar squad preparation of the round, and the manipulation and firing of the mortar.

B-4. The procedures to fire the FRPC are the same as for the M934 service cartridge. A major advantage of using the M931 FRPC is the elimination of the time required to retrieve and clean training devices. It also does not exceed 75 percent of the M934 cartridge's unit production cost.

Note. The FRPC is treated as a regular service mortar cartridge. It has the same range and firing restrictions as a service round.

B-5. The M931 consists of the following major components:
- Projectile body assembly.
- M233 propelling charges (with M47 propellant).
- M1005 ignition cartridge (with M44 propellant).
- M34 fin assembly.
- M781 point detonating practice fuze.

This page intentionally left blank

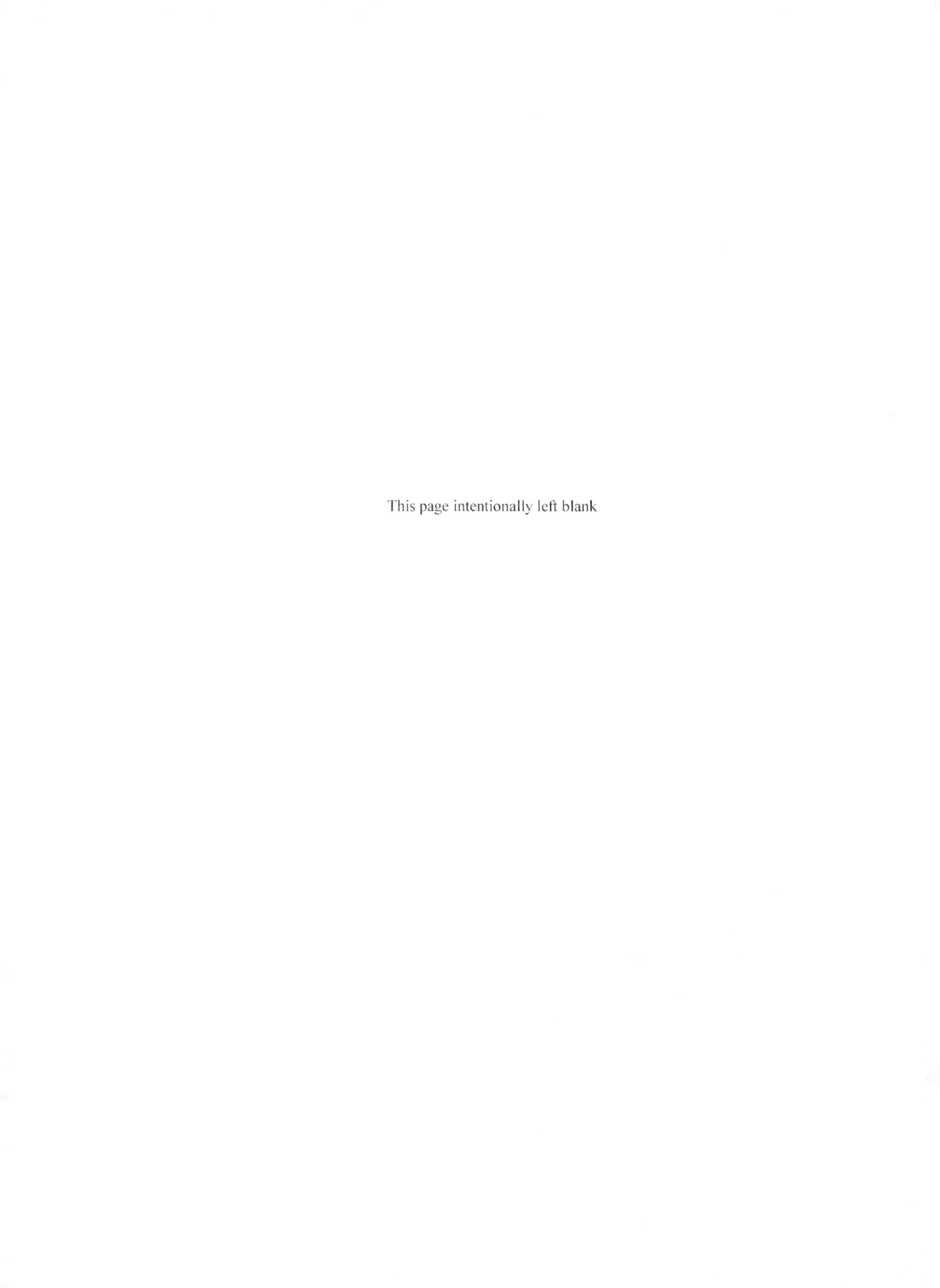

Glossary

The glossary lists acronyms and terms with Army or joint definitions. Where Army and joint definitions differ, (Army) precedes the definition. Terms for which TC 3-22.90 is the proponent are marked with an asterisk. The proponent manual for other terms is listed in parentheses after the definition.

Acronym	Definition
ABCT	Armored brigade combat team
ASIP	Advanced System Improvement Program
ASP	ammunition supply point
AZ	azimuth
BAD	blast attenuator device
CFF	call for fire
CI	commander's interface
DAGR	defense advanced GPS receiver
DAP	distant aiming point
DLY	delay
EL	elevation
EOD	explosive ordnance disposal
EPDA	enhanced power distribution assembly
FBCB2	Force XXI Battle Command, Brigade-and-Below
FCC	fire control computer
FDC	fire direction center
FFE	fire for effect
FIST	fire support team
FO	forward observer
FP	firing point
FPF	final protective fire
FRPC	full range practice cartridge
FT	firing table
GD	gunner's display
GPS	global positioning system
G-T	gun-target
H^3	radioactive tritium gas

Glossary

HE	high-explosive
HMMWV	high mobility multipurpose wheeled vehicle
HOB	height of burst
IBCT	Infantry Brigade Combat Team
ICV	Infantry carrier vehicle
ILLUM	illumination
IMP	impact
IMU	inertial measurement unit
IR	infrared
LARS	left add/right subtract
LCD	liquid crystal display
LHMBC	lightweight handhed mortar ballistic computer
MFCS	Mortar Fire Control Systen
mm	millimeter
MTSQ	mechanical time, superquick
NATO	North Atlantic Treaty Organization
NAV	navigation
NCO	noncommissioned officer
NSB	near-surface burst
OIC	officer in charge
OpMov	operationally moving
O-T	observer-target
PD	point detonation
PDA	power distribution assembly
PDMA-DQR	pointing device mount assembly-dynamic quick release
PLGR	precision lightweight GPS receiver
PRX	proximity
PUBS	portable universal battery supply
RM	risk management
RSTA	reconnaissance, surveillance, targetting, and acquisition
RTO	radiotelephone operator
SBCT	Stryker brigade combat team
SDZ	surface danger zone
SOP	standard operating procedure

SQ	superquick
VMS	vehicle motion sensor
WP	white phosporuous

SECTION II – TERMS

This section has no entries.

This page intentionally left blank

References

REQUIRED PUBLICATIONS

ADRP 1-02, *Terms and Military Symbols*, 16 November 2016.

Department of Defense Dictionary of Military and Associated Terms, 15 December 2016.

RELATED PUBLICATIONS

Most Army doctrinal publications and regulations are available at: http://www.apd.army.mil.

Most joint publications are available online at: http://www.dtic.mil/doctrine/doctrine/doctrine.htm.

Other publications are available on the Central Army Registry on the Army Training Network, https://atiam.train.army.mil.

Mortar firing tables are available through the U.S. Army Research, Development and Engineering Command (RDECOM), Armament Research, Development and Engineering Center (ARDEC) who controls the publications management application (PMA) of Firing Tables and Ballistics: https://picac2as2.pica.army.mil/login/pma.htm

AR 385-63/MCO3570.1C, *Range Safety*, 30 January 2012.

ATP 5-19, *Risk Management*, 14 April 2014.

ATTP 3-21.90, *Tactical Employment of Mortars*, 4 April 2011.

DA PAM 350-38, *Standards in Training Commission*, 22 November 2016.

DA PAM 385-63, *Range Safety*, 16 April 2014.

FM 3-22.91, *Mortar Fire Direction Center Procedures*, 17 July 2008.

FM 27-10, *The Law of Land Warfare*, 18 July 1956.

TC 4-02.1, *First Aid*, 5 August 2016.

TM 9-1010-223-10, *Technical Manual Operator's Manual For Lightweight Company Mortar, 60mm: M224 (NSN 1010-01-020-5626)*, 1 October 2012.

TM 9-1010-233-10, *Operator's Manual for Lightweight Company Mortar, 60-mm, M224A1 (NSN 1010-01-586-2874)(EIC:4SC)*, 1 July 2016.

TM 9-1010-233-23&P, *Technical Manual Field Maintenance Manual Including Repair Parts and Special Tools List (Including Depot Repair Parts) for Mortar, 60-mm, M224A1 (Lightweight Company) (NSN 1010-01-586-2874) (EIC:4SC)*, 1 September 2016.

TM 9-1015-249-10, *Operator's Manual for Mortar, 81-mm, M252 (NSN: 1015-01-164-6651) (EIC:4SK)*, 10 August 2012.

TM 9-1015-250-10, *Technical Manual Operator's Manual for Mortar, 120mm: Ground Mounted M120 (NSN 1015-01-226-1672) (EIC:4SL); Mortar, 120MM: Carrier-Mounted (10-15-01-292-3801) (EIC:4SE)*, 30 April 2012.

TM 9-1015-256-13&P, *Operator and Field Manual Including Repair Parts and Special Tools List (Including Depot Repair Parts) for Mortar, 120mm, M120A1 (NSN: 1015-01-554-0749) (EIC:4SN)*, 31 August 2012.

References

TM 9-1015-257-10, *Operator's Manual for Mortar, 81-mm, M252A1 (NSN: 1015-01-586-2135) (EIC:4SU)*, 25 August 2014.

TM 9-1015-257-23&P, *Field Maintenance Manual Including Repair Parts and Special Tools List for 81-mm Mortar, M252A1 (NSN: 1015-01-586-2135) (EIC:4SU)*, 25 August 2014.

TM 9-1220-248-10, *Operator's Manual for Mortar Fire Control System, M95 (NSN: 1230-01-503-7784) (EIC:3QT) (With Version 3 Software)*, 31 October 2005.

TM 9-1230-205-10, *Technical Manual Operator's Manual for Mortar Fire Control System-Dismounted 120mm, M150 (NSN 1230-01-560-1027) (EIC:3QT); Mortar Fire Control System-Dismounted 120mm, M151 (NSN 1230-01-560-1028) (EIC:3QS) (Hardware Only)*, 30 October 2013.

TM 9-1290-333-15, *Operator's, Organizational, Direct Support, General Support and Depot Maintenance Manual (Including Repair Parts and Special Tools List): Compass, Magnetic, Unmounted: M2*, 7 November 1963.

TM 9-2350-277-10, *Technical Manual Operator's Manual for Carrier, Personnel, Full-Tracked, Armored, M113A3 (NSN 2350-01-219-7577) (EIC:AEY); Carrier, Command Post, Light Tracked, M577A3 (NSN 2350-01-369-6085) (EIC:AE7); Carrier, Standardized Integrated Command Post System (SICPS) M1068A (NSN 2350-01-369-6086) (AFC); Carrier, Mechanized Smoke Obscurant M58 (NSN 2350-01-418-6654) (EIC:5CG)*, 15 March 2012.

TM 9-2355-311-10-3-1, *Operator's Manual Volume 1 of 3 Mortar Carrier Vehicle (MCV) M1129A1 (NSN: 2355-01-505-0871) (DIC:AFJ) Stryker*, 30 September 2016.

TM 9-2355-311-10-3-2, *Operator's Manual Volume 2 of 3 Mortar Carrier Vehicle (MCV) M1129A1 (NSN: 2355-01-505-0871) (EIC:AFJ) Stryker*, 30 September 2016.

TM 9-2355-311-10-3-3, *Operator's Manual Volume 3 of 3 Mortar Carrier Vehicle (MCV) M1129A1 (NSN: 2355-01-505-0871) (EIC:AFJ) Stryker*, 30 September 2016.

TM 9-6675-262-10, Technical Manual *Operator's Manual for Aiming Circle, M2 W/E (NSN 1290-00-614-0008) and M2A2 W/E (6675-01-067-0687)*, 6 December 2013.

PRESCRIBED FORMS

Unless otherwise indicated, DA forms are available on the Army Publishing Directorate (APD) web site (www.apd.army.mil).

DA Form 5964, *Gunner's Examination Scorecard – Mortars*.

REFERENCED FORMS

Unless otherwise indicated, DA forms are available on the Army Publishing Directorate (APD) web site (www.apd.army.mil).

DA Form 2028, *Recommended Changes to Publications and Blank Forms*.

DA Form 2188-1, *LHMBC/MFCS Data Sheet*.

DA Form 2408-4, *Weapon Record Data*.

Index

Entries are listed by paragraph numbers unless mentioned otherwise.

2

25-meter sight box method. 2-17

8

81-mm mortar
   dismounting. 4-34
   loading. 4-33

A

aiming circle, 1-13, 6-6
aiming circle methods. 2-79
aiming posts. 5-24, 2-69
   lights. 2-71
   placing. 2-83
   realigning. 4-32, 3-34
ammunition
   120-mm. A-31
   60-mm. A-22
   81-mm. A-27
   color codes, A-4
   field storage. A-18
   lot number. A-13
arm-and-hand signals. 1-86

B

baseplate. 4-12
bipod. 4-11
bracketing method. 7-42
buffer zone. 1-58

C

carrier-mounted 120-mm mortar
   deflection and elevation. 5-65
charge. 1-82
cook off. 3-38
creeping method. 7-44

D

DA Form 5964, Gunner's Examination Scorecard–Mortars, 8-3
deflection. 2-85, 2-84, 1-81
   changes. 4-30, 3-33
   setting. 2-20
direct-lay method steps. 7-11
dismount and carry the mortar. 3-47
dispersion area. 1-57
distant aiming point method. 2-15

E

elevating mechanism. 4-14
elevation. 1-84
   changes. 4-30, 3-33
elevation setting. 2-19
emplace the electronic rack. 5-39
error report. 1-98
examining board. 8-7
   general rules. 8-12

F

fire commands. 7-26, 6-70, 1-67
   modified. 7-32
fire direction center (FDC), 5-10, 6-9
fire support team. 1-5
   fire direction center. 1-10
   forward observer. 1-7
   mortar squad. 1-12
firing position. 1-53
FPF mission, 6-24
full-range practice cartridge. B-1
fuze wrench, M18. A-88
fuzes. A-36, A-11
   120-mm M734. A-52
   120-mm M734A1. A-63
   120-mm M935. A-71
   60-mm M734. A-47
   60-mm M734A1. A-55
   60-mm M935. A-67
   81-mm M524-series. A-74
   81-mm M526-series. A-76
   81-mm M734. A-50
   81-mm M734A1. A-59
   81-mm M935. A-70
   M567. A-77
   M745. A-80
   M751 practice. A-89
   M772. A-84
   M775 practice. A-89, A-83
   M776. A-91, A-82
   M779. A-83
   M783, A-26
   M84. A-85
   M84A1. A-87
   proximity. A-37

G

ground-mounted 120-mm mortar. 5-19
   deflection and elevation changes. 5-22
   loading and firing. 5-25
   place into action, 5-34

Index

safety checks. 5-21
take out of action. 5-26
ground-mounted mortar examination test. 8-22
gunner's quadrant. 2-23

H

hangfire. 3-37

I

illumination. 7-60
impact area. 1-55

L

ladder method. 7-46
laying for direction. 2-58

M

M115 boresight. 2-9
M121 carrier-mounted mortar examination. 8-49
M150 emplacement. 5-35
M170 bipod assembly. 3-9
M177 mount. 4-10
M191 bipod assembly. 5-13
M2 compass operation. 2-32
M225 cannon assembly. 3-8
M253 cannon assembly. 4-9
M298 cannon assembly. 5-10
M326 mortar stowage kit
   basic issue items box. 5-31

M45-series boresight. 2-4
M67 elbow telescope. 2-42
M7/M7A1 baseplate. 3-16
M8/M8A1 baseplate. 3-17
M95 MFCS
   dismounted. 6-28
   mounted. 6-5
M9A1 baseplate. 5-14
manual boresight setup. 5-107, 5-100
mask and overhead clearance. 5-63, 4-24, 3-28
mortar sights. 2-81
mortar squad. 1-12
   communication. 1-100

N

natural object method. 7-24
night fire. 1-105

O

operating without an FDC. 7-4

P

plotting boards. 2-73

R

risk management assessment. 1-16
RMS6-L 120-mm mortar system. 5-90
RMS6-L Stryker-mounted mortar examination. 8-49

S

safety "T". 1-34
safety checks. 4-22, 3-26
safety officer. 1-27
   checklist, 1-5
searching fire. 7-63, 1-78
section right/left. 1-73
setting deflection. 2-50
setting elevation. 2-54
short round. 3-39
sight calibration, 2-4
sight picture. 2-57
sight unit maintenance. 2-63
smoke. 7-59
spotting. 7-50
Stryker mortar vehicle. 5-75
   configurations. 5-87
   mortar ammunition storage. 5-93
   squad. 5-77
   weapons system. 5-82
surface danger zone. 1-51

T

target engagement. 1-65
traversing fire. 7-64, 1-77
traversing mechanism. 4-13

V

volley fire. 1-72

TC 3-22.90
17 March 2017

By Order of the Secretary of the Army:

MARK A. MILLEY
General, United States Army
Chief of Staff

Official:

GERALD B. O'KEEFE
Administrative Assistant to the
Secretary of the Army
1705303

DISTRIBUTION:
Active Army, Army National Guard, and United States Army Reserve: To be distributed in accordance with the initial distrubution number (IDN) 116075, requirements for TC 3-22.90.

PIN: 201470-000